TESS WHITEHURST

The Good Energy Book

"Like feng shui for the soul, this book takes you step-by-step through how to clear cluttered energy, keep the good stuff flowing—and transform your life in the process!"

—Sara Wiseman, author of
Writing the Divine and *Your Psychic Child*

"A classic and splendid book about natural self-healing for the home, body, and spirit. It's truly inspiring and beautifully written. Read it and revitalize your own personal space and intuitive eye!"

—Jodi Livon, author of
The Happy Medium

"Sparkling advice and simple techniques for creating and maintaining happiness and harmony at home, at work, and in the world. For practical guidance on cultivating positive energy, Tess is the best!"

—Annie Wilder, author of
House of Spirits and Whispers

"Tess Whitehurst's book is clear and concise, offering you everything you need to know about creating good energy for yourself and your home. Whether you consider yourself a beginner on this subject or not, this informative and inspiring reference book belongs in your library."

—Terah Kathryn Collins, author of
The Western Guide to Feng Shui and
The Western Guide to Feng Shui for Prosperity

Magical Housekeeping

"Filled with valuable information and ancient wisdom to activate sparkling energy and create true sacred space in your home. I recommend it!"

—Denise Linn, author of *Sacred Space*

"This is a magical book! It will change the way you think about even the most mundane housekeeping concerns. Infinitely useful and fun, and full of exciting and accessible tips, this book reminds us that even the seemingly simple things in your life are inroads to the most profound. If you are ready for positive change, you can't do better than this! It makes us smile every time we think of this book and of the joy it will bring into people's lives."

—Ana Brett & Ravi Singh,
internationally respected yoga teachers

"Tess Whitehurst shows us that cleaning can be a powerful tool for personal transformation. *Magical Housekeeping* is a blend of New Age techniques that incorporates the metaphysical topics of mantras, angels, fairies, crystals, and plant and animal allies. Get your brooms ready ... housekeeping will never be the same again."

—Ellen Dugan, author of *Cottage Witchery*,
Natural Witchery, and *Garden Witch's Herbal*

the GOOD ENERGY BOOK

Tess Whitehurst is an intuitive counselor, energy worker, feng shui consultant, and speaker. She has appeared on the Bravo TV show *Flipping Out*, and her writing has been featured in such places as the AOL welcome page, Llewellyn's annuals, and the *Whole Life Times* blog. To learn about her workshops, writings, and appearances, and to sign up for her free monthly newsletter, visit her online at www.tesswhitehurst.com.

Tess Whitehurst

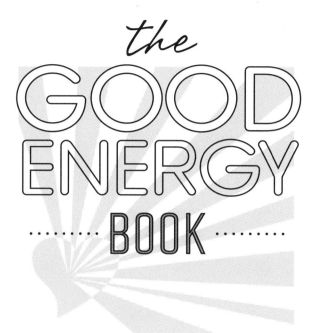
the GOOD ENERGY BOOK

Creating Harmony and Balance
for Yourself and Your Home

Llewellyn Publications
Woodbury, Minnesota

FIRST EDITION
Seventh Printing, 2016

Book design and edit by Rebecca Zins
Cover credits: flower, ©PhotoDisc;
background, ©iStockphoto.com/Julia Kaptelova
Cover design by Ellen Lawson
Interior: heart/ray ornament from the Fleurons of Hope font, a 2005 Font Aid III project. Uniting the typographic and design communities, Font Aid III raised funds to expedite relief efforts in countries affected by the Indian Ocean earthquake and tsunamis. More than 220 designers worldwide submitted over 400 glyphs for the collaborative Fleurons of Hope typeface.

Llewellyn is a registered trademark
of Llewellyn Worldwide Ltd.

Library of Congress Cataloging-in-Publication Data
Whitehurst, Tess, 1977–
  The good energy book / Tess Whitehurst.—1st ed.
      p. cm.
  Includes bibliographical references (p.   ) and index.
  ISBN 978-0-7387-2772-1
  1. Spiritual life—Miscellanea. 2. Force and energy—Miscellanea. I. Title.
  BF1999.W5365 2012
  131—dc23
                                                                2011035674

Llewellyn Worldwide Ltd. does not participate in, endorse, or have any authority or responsibility concerning private business transactions between our authors and the public.
  All mail addressed to the author is forwarded, but the publisher cannot, unless specifically instructed by the author, give out an address or phone number.
  Any Internet references contained in this work are current at publication time, but the publisher cannot guarantee that a specific location will continue to be maintained. Please refer to the publisher's website for links to authors' websites and other sources.

Llewellyn Publications
A Division of Llewellyn Worldwide Ltd.
2143 Wooddale Drive
Woodbury, MN 55125-2989
www.llewellyn.com
Printed in the United States of America

contents

CONTENTS

six

Contents

INTRODUCTION

rom a very young age, most of us are taught to believe something that isn't true. In fact, we're taught to believe it so intrinsically that we don't even realize we believe it—we think we *know* it. It's a belief that insidiously works its way under and behind all our thoughts and experiences, undermining our instincts, divorcing us from meaning, separating us from a direct experience of the present moment, and depriving us of our divinely given spiritual abilities.

What is this belief?

The belief in separation: separation between the seen and the unseen, between form and spirit, between past and future, between self and other, and between what we scientifically "know" and what we spiritually sense.

After many years of the depression, disempowerment, and debilitating fear that are the inevitable results of this belief, I began working with magical and metaphysical modalities such as meditation, crystal healing, space clearing, feng shui, and ritual.

Over time, through direct experience and observation, I came to understand the following:

- There is no separation between the seen and the unseen, form and spirit, and the known and the unknown.
- In truth, everything is connected, and everything affects and interacts with everything else.
- When we tune in to our intuition and embrace the connection between the seen and the unseen, we can absolutely, positively affect our life situation and create the life conditions we want.

In this book, I share the healing and clearing practices that have helped to transform my life from a near-constant experience of fear and struggle to a near-constant experience of adventure and delight. It is my heartfelt wish that you experience the same.

Brightest blessings,

Tess

one

Energy and the Subtle Reality

*I*n truth, there is only love. Yes, there is the appearance of duality—light and dark, happy and sad, seen and unseen, known and unknown—but behind and inside of all of these appearances is a perfect, whole, unified field of energy. When we can tap into this field of energy (i.e., stay connected to it even when, on a surface level, we may perceive the illusion of discord), we can consciously direct it toward healing ourselves, healing others, and infusing our lives with ever-greater levels of harmony, balance, and peace.

You've opened this book because on some level, you already sense this unified field of love that transcends and underlies "everyday reality." What's more, you want to know more about how to access it to shift things for the better: to transmute negativity into positivity and leave beautiful, sparkling energetic conditions in your wake.

And you've come to the right place.

Initiatory Invocation
Embrace the Mystery

Let's face it: everything is a mystery. And, as much as we might learn about physics or philosophy or religion (or anything else), it seems there is simply no way to truly unravel it.

Many people have a hard time with this. Whether they lean toward the science camp or the religious camp (or both), there are tons of folks feverishly trying to map out exactly what is and is not really going on here.

While there are valid and valuable aspects of both science and spirituality, as an energy worker and a student of the subtle realm, you are called to stand facing the mystery head-on, with your feet planted firmly in the mystery and with an awareness that the mystery is surrounding you, within you, and towering above you. Why? First, because it's the only honest thing to do. Second, because this mystery-embracing practice (which is ongoing) will allow you to be present with what *is*, rather than projecting an idea onto the world of what is *supposed* to be. This, in turn, will open the floodgates to a level of perception that is not only uncommonly deep and satisfying but that is a prerequisite to sensing and working with energy in an effective and beneficial way.

Here's your first assignment. Grab a notebook and a pen. Relax, take a few deep breaths, and consider your current concepts about reality and the nature of existence. Really get clear on things you formerly may have taken for granted, such as externally imposed theoretical structures or ways you've come to almost imperceptibly "explain away" the mystery. In other words, are you holding on to any beliefs or concepts that may *conceivably* be inaccurate? Any that you perhaps picked up along the way and adopted without really questioning them? You might even discover that when you really look at them, you find some of these formerly unquestioned ideas to be absurd or childlike. (This is natural and nothing to be ashamed of.)

Here are just a few examples of the types of concepts I'm talking about:

- Time is linear. Like a line, it has a beginning, a middle, and an end.

- Reality can be described by the adjectives *cold* and *hard*.

- Sane people can all agree on what's real and what's not real.

- Someone, somewhere, has it all figured out.

- God exists.

- God doesn't exist.

- God has a beard (or is a woman, or wears a robe, or hates it when people have sex, etc.).

Once you've accessed these concepts, brought them into the light of awareness, and made a list of anything that could possibly be untrue or at least questionable, take a moment with each one and see if you can conceive of ideas that contradict it. Have a mental conversation with yourself and really do your best to disprove or at least partially discredit each notion.

For example, consider the concept that time is linear and has a beginning, a middle, and an end. After writing down this concept, you might question it by writing in response, "No one has ever proved to me that time is linear—it could be more like a wheel or a figure 8 for all I know. Theoretical physicists like Stephen Hawking and Albert Einstein say that time is relative and is experienced differently depending on where you are and how fast you are moving. It's said that if we could enter black holes without dying we could conceivably travel through time, which would indicate that time is not linear at all."

When this feels relatively complete with each statement (for the time being), take a deep breath, close your eyes, and simply be willing to let go of fixed concepts about things that are ultimately mysterious. Be willing to enter into the mystery. To help with this willingness, affirm inwardly or aloud: "I am willing to accept the mystery. I am willing to surrender to the mystery. I am willing to dwell in the mystery."

As an initiatory exercise, you might even like to take a walk in nature or watch a sunrise or sunset while leisurely contemplating each part of this affirmation. This will help the process of stripping away the old so that a constant, conscious awareness of the mystery can begin to take over and settle into your consciousness.

For the duration of reading and working with the practices in this book, it's a good idea to say this affirmation immediately upon waking. As a reminder, you might even write it on a note card, frame it, and place it on your nightstand.

As I mentioned above, ideally this practice of letting go of rigid, outmoded concepts about the nature of time and reality will be ongoing. The more you release, the more connected you will be to the subtle realm (which includes the realm of infinite love, and also the many layers of feelings and invisible energies that characterize and project the qualities of our present life experience) and the more spiritually powerful you will be. In time, you will find yourself waking up every day to an overwhelming feeling of awe at the nature of life and existence. "Another day here in this mysterious, beautiful world!" you'll think; "How fun!" And this will create space and open doors of possibility in your life like nobody's business.

And please note that it's not my intention for you to take anything I've written to be absolute truth. After all, while I am using words to write this book, the subtle realm is beyond words—so, at best, when I'm talking about the subtle realm, I can only approximate or point to what I am trying to communicate. And even then, I can only talk about my own perceptions and conclusions, not yours. It's up to you to experience the subtle realm firsthand, to learn experientially, to come to your own conclusions, and to formulate your own theories and techniques that work best for you.

More About Perceiving the Subtle Realm

To clarify what I mean by *subtle realm*, you might think of it as an umbrella term for everything that is beneath what we, in our culture, call "everyday reality." At its very base, this is a pure, open, unified, vast, conscious presence we may call love. Above that, there are denser but still invisible energies that begin to take on and characterize the illusion of duality (bright/dark, light/heavy, etc.). And, although duality is an illusion, it is an illusion that is very pervasive in this life experience that we are in the middle of, so it can be very useful to consciously work with this level of subtle reality to effect the types of positive change we desire.

To orient you further to what I mean by the subtle realm, I want to let you know that whether we consider ourselves psychic or intuitive (or not), every one of us already has a sense of it. For example, we all immediately understand expressions like "shady character," "electric personality," and "heavy subject." The character is usually not standing in the shade; neither is the personality plugged into a light socket or the subject made of lead. This is because all of these expressions aren't talking about physical reality but about subtle, or energetic, reality. In other words, we're speaking about something invisible but nonetheless real and recognizable.

When you begin to consciously cultivate an awareness of the subtle realm, you might first become aware that there are rooms and areas that make you feel happy or light or carefree, and rooms and areas that make you feel sad or heavy or worried. You might also become aware that some people seem to have a vibration that attracts you and some people seem to have a vibration that repels you. And you might feel impelled to approach—or steer clear of—certain objects and situations based on your premonitions and energetic perceptions.

Then, as you get even deeper into this awareness, you will not only begin to sense patterns and recognizable formations in the energy, but you will become aware that you can interact with, shift, and even transform these patterns and formations. And that brings us to…

THE PRACTICES YOU'RE ABOUT TO LEARN

To reiterate, everything is one unified field of energy. And, when we become aware of the invisible layers of the appearance of our present life experience, we start to notice that some things appear to have "good energy" and some things appear to have "bad energy." However, in reality, energy is neutral, although it may presently seem to appear as trapped, stagnant, and festering or flowing, healthy, and vibrant. Trapped energy can be transmuted into vibrant energy, but while it's trapped it can (and does) often appear to be bad and attract negative conditions, feelings, and energetic patterns; conversely, vibrant energy attracts positive conditions and energetic patterns. You might compare energy to water or air. It is not good or bad in and of itself, but when it becomes stagnant it can be a breeding ground for unhealthy conditions.

Just as swamp water can attract and foster organisms that make it dangerous to drink, trapped energy can attract negative emotions and toxic conditions. The good news is that we can engage in practices that get the energy moving in a healthy way and transform trapped energy into vibrant energy. This not only enhances our spaces, spirits, and life conditions with positivity and sparkle, it also infuses us with a feeling of empowerment and self-mastery, and (since everything is connected) also helps heal and transmute the energy of the entire planet.

While you'll learn about a whole spectrum of these practices in this book—including energetic shielding, blessing, and fine-tuning—I've found that when it comes to being happy and manifesting positive conditions in our lives, the most important considerations are (a) clearing our own personal energy fields and (b) clearing the energy fields of our homes.

To illustrate the invisible mechanics of energy clearing, let's use the water metaphor again. Imagine a waterlogged stone fountain that has been sitting stagnant for some time. Somewhere along the way, various plant and bacteria organisms found their way into the water, and it began to turn green and stink. Then imagine that a gardener came along and transferred the water to a lake where the bacteria and plant organisms could thrive in a spacious and diverse habitat. Next, he cleaned the bottom of the fountain thoroughly with natural, organic cleaners and replaced the water with fresh, clear, sparkling well water. Finally, he turned on the fountain so that it flowed, sparkled, generated negative ions (the scientific term for good vibes), and stayed fresh.

Metaphorically speaking, the fountain is your home (although it could just as easily be you). The water is the energy in your home. The bacteria and plant organisms are trapped entities, emotions, and thought forms. The lake is the infinite, unified field of love that underlies all things in truth, which always brings order and orchestration to all other layers of perception, and which we can always access simply by becoming conscious of it. And the gardener (the spiritual healer or practitioner, i.e., *you*), instead of using actual cleaning supplies (though those may also come in handy), clears the energy with his focused attention helped along by things like sound (such as chimes and rattles), prayer, chanting, and natural ingredients (such as sage smoke and sea salt).

WHY THE PRACTICES IN
THIS BOOK ARE HELPFUL

For most of the period that humans have been alive on this planet, we've spent the bulk of our time outdoors. And, still speaking macrocosmically, in our not-so-distant past, our dwellings were less permanent and closer to the wildness of nature. This means that a natural energetic cleansing system was in place. Stagnant energy—as well as energetic residue from old thoughts, words, and emotional patterns—in and around our physical bodies and our living spaces was (generally speaking) not long for this world, because wind would move it, sunlight would activate it, rain would wash it, and the earth would absorb it.

Of course magical and shamanic practices were still employed on occasion to embody and channel the omnipresent current of divine natural energy in order to cleanse and heal spaces and people. But now that we, as a culture, have moved indoors and made a fundamental split from the natural world, those of us who are energetically sensitive must employ similar practices on a regular basis if we want to insure that our bodies, lives, and living spaces are consistently characterized by clear and sparkling energy. By the same token, studying and employing energy-healing modalities powerfully realigns us—individually and as a culture—with the natural world, which is of great significance at this time in history.

two

PRELIMINARY PREPARATIONS: THE INITIAL SHIFT

he field of pure love that underlies everything wants to pour through your life experience like wind wants to blow and waves want to eternally wash upon the shore. The purpose of this chapter is to clear away any illusion-based blocks to this natural, positive momentum so that joy can blossom and blessings can flow. We'll accomplish this by doing an initial sweep of negative (stagnant) energy to establish a clearer and more buoyant vibration in your home. In turn—because everything is connected—this will begin to shift your personal energy in ways that will clear your mind, amplify your power, and prepare you for the exciting journey ahead.

CLEAR CLUTTER

No matter where it is or how often you consciously think about it, you have a relationship with everything you own. In other words, you must have at least one or two thoughts, feelings, memories, or obligations attached to something, otherwise you wouldn't be able to recognize it as yours. So—in this unified sea of energy that we call existence—you might imagine that (energetically speaking) your relationship with each item you own is a cord of energy. If you love, use, or need something, that item sends you positive energy through that cord. (To illustrate, think of the subtle charge you get from your favorite pair of shoes or your freshly cleaned, whisper-soft organic cotton sheets.) If you *don't* love, use, or need something, that item either drains energy from you or sends negative energy your way. (Now think of the slight yet significant energy drain you feel when you wade through a closet full of clothes you don't like or dust that ugly porcelain rooster that was a gift from your in-laws.) With this in mind, you can see that, in essence, every piece of clutter that you own is an actual energy leak. Over time and especially considering all the many things you own, this can add up to a very significant drain on your vitality and overall well-being.

What's more, everything is connected. There is an inextricable link between the subtle realm and the physical realm, just as there is an inextricable link between the items in your home and your thoughts, feelings, beliefs, and life conditions. To rephrase this idea: when you clear clutter from your home, you're simultaneously clearing clutter from your life. What's more, you're enhancing and fine-tuning the vibrations of the subtle realm, making it less likely that unwanted conditions, gloomy feelings, household discord, and general negativity will hang around. In turn, your home will be happier, and your future energetic endeavors will be more grounded, organic, and effective. (To illustrate: if you're a Harry Potter fan, consider how much drama could have been avoided if Hogwarts had stayed on top of the clutter clearing!)

So go through your drawers and closets, the trunk of your car, the attic, the basement, your rented storage space, and anywhere else you might be harboring items that you don't love, use, or need. Take as long as you need with this, no matter if it takes a week, a month, or a year. If you feel overwhelmed by the prospect, commit to clearing out a single dresser or even simply a drawer. Then, later, repeat this with another small area until you naturally begin to gain momentum.

As you clear clutter, keep a bottle of fresh water nearby and drink, drink, drink. Clearing clutter is a process of releasing and flushing energetic toxins from your mind, body, and emotions, so the water (in addition to benefitting your health and keeping you hydrated) will help this process along.

The following checklist (excerpted from *Magical Housekeeping*) can help you recognize clutter when you see it:

TABLE 2-1: **Clutter-Clearing Checklist**

PAPER	Old receipts Old warranties and other unnecessary documents Junk mail Old cards and love letters Expired coupons
CLOTHES	Clothes that don't fit Clothes you don't love Clothes that make you feel unattractive or less than stunning Clothes you never wear Clothes that need to be repaired that you know you'll never repair
BOOKS	Any book you'll never open again
DECORATIONS	Any decoration that doesn't uplift you or bring you joy Any image depicting a condition or feeling you don't want to experience Dried or faux flowers or plants that appear faded, dusty, brittle, or overly dead
FURNITURE	Any piece that doesn't fit in your house Any bed, couch, or dining table you shared with an ex-partner Anything you don't love Pieces that injure, trip, or inconvenience you

GIFTS	Anything you're hanging onto out of guilt or obligation
FOOD	Anything you're honestly never going to eat
CAR CLUTTER	Trash Anything that doesn't belong in your car
UNFINISHED PROJECTS	Anything you're not going to (honestly) finish in the next month
BROKEN THINGS	Any broken item you can't or aren't willing to fix (unless it's still useful and convenient and you honestly don't mind that it's a little bit broken)
ITEMS WITH NEGATIVE ASSOCIATIONS	Gifts or hand-me-downs from people with whom you have negative associations Anything that reminds you of a negative situation or period

Note: If this section hasn't sufficiently sparked your clutter-clearing engines, please see chapter 1 of *Magical Housekeeping: Simple Charms and Practical Tips for Creating a Harmonious Home* or my mini e-book *Magical Clutter-Clearing Boot Camp*.

CLEAN

Since what we call negative energy is often simply stagnant energy, nothing gets the energy flowing in the right direction like a thorough physical cleaning. Furthermore, intention directs energy like a laser beam. So when you clean with the strong intention to clear negativity and lift the vibe, guess what? *You clear negativity and lift the vibe.* So get to work! Don't go too crazy (unless you want to), but push yourself a bit more than you usually do when you clean. For example, you might clean under and behind things that normally stay put, steam clean the curtains, shake out the area rugs, flip the mattress, etc.

And there are a number of natural ingredients—such as essential oils, vibrational essences, and even pure thought energy—that you can add to your cleaning supplies to enhance the effect. Here are some ideas.

Essential Oils

As you may know, an essential oil is a highly fragrant, highly concentrated compound of a particular plant and is available online and at most health food stores. Some essential oils can irritate the skin, so be careful when handling them. The essential oils in the list below have highly spiritual vibrations and can significantly help lift and purify the vibration of your home. Start with a 100 percent natural biodegradable unscented cleaner and then consider adding ten to twenty drops of any of the following oils in any combination:

- Cedar
- Lemon

- Clary sage

- Sage

- Lavender

- Rosemary

- Tea tree

- Eucalyptus

- Spruce

Vibrational Essences

Vibrational (flower and gem) essences, unlike essential oils, are not fragrant. Rather, they're the spiritual vibration of a blossom or mineral preserved in water and brandy. Any of the following can be added to natural cleaners or air fresheners to impart a fresh and positive vibration in the space.

Note: Essences are also available online and at many health food stores. You can also easily learn how to make your own. The book *Natural Healing Wisdom & Know-How*, compiled by Amy Rost, has a good description of how to create your own flower essences, and you can learn how to make gem essences by taking a look at the gemstones chapter of *Magical Housekeeping*.

FLOWER ESSENCES

- Bach Rescue Remedy

- Crab apple essence

- White chestnut essence

- Rose essence

GEM ESSENCES

- Fluorite essence
- White quartz essence
- Rose quartz essence
- Citrine essence

Thought Energy

You may also just as effectively empower your cleaning solutions and natural air fresheners with your thoughts, prayers, visualizations, and intentions. For example, you might engage in any of the following mini blessing rituals:

- Hold your cleaning solution in both hands and close your eyes. Say something like this: "God/Goddess/All That Is, please infuse this cleaner with vibrations of purity, clarity, and love." Then visualize very bright white light coming down from above, entering the crown of your head, going down to your heart, down your arms, and out through the palms of your hands and into the bottle. Visualize/imagine/feel the bottle swirling and pulsating with this light.

- Set your cleaning solution in bright sunlight and say or think: "I call on the light of the sun to activate and enliven the purifying and clarifying aspects of this liquid." Allow it to continue to soak up the sun for three to five minutes.

MAKE NOISE

Now we're going to engage in some space-clearing techniques. In other words, we're going to work in the subtle energetic realm to clear the air and shift the vibe for the better.

Sound can get energy moving in a healthy way by loosening and unsticking negative, stagnant energy and helping transmute it into positive, vibrant energy. This is why so many shamanic healers and energy workers treasure their rattles and chimes. In fact, a rattle alone (or even just clapping your hands—see below) can often be all that's needed to sufficiently clear the energy in a space. Which brings us to…

The Perfect Space-Clearing Rattle

I used to just clap loudly around the perimeter of each room and area in my home to purify with sound. And while this also works, after I was left with red, aching hands one too many times, the instructions for making this rattle began to appear in my mind, and I quickly acted on them. If you decide to follow suit, you'll find that—in addition to being perfect for space-clearing purposes—this rattle is also remarkably easy and inexpensive to make. And, since you're making it yourself, you'll have an immediate synergistic affinity with it; it will be like an extension of you and your personally channeled power, which is always ideal when it comes to shamanic or spiritual tools.

YOU WILL NEED:

1 large metal saltshaker with handle

2 small handfuls dry white beans

2 small handfuls dry black beans

1 clove garlic

Sunlight or white sage

First, purify the saltshaker by immersing it in sunlight or white sage smoke for one to two minutes. Then hold one small handful of white beans in your right hand. Cradle your right hand in your left hand, and visualize/imagine/feel the beans glowing, swirling, and pulsating with very bright white light. Mentally empower them with the intention to help purify the space of all negative and stagnant energy. Similarly empower the clove of garlic with light and intention, then cut it into eight pieces, and place the pieces in the shaker with the rest of the ingredients. Repeat this process with one more handful of white beans and two handfuls of black beans. Close the lid, hold the rattle in both hands, and consecrate the shaker to the Great Spirit, God/Goddess, Universe, or whatever name you use for the Divine by holding it up and visualizing very bright white light descending from above and filling the rattle. You might say a prayer or invocation such as:

> *Great Spirit (or Goddess, Divine Presence, etc.),*
> *I now consecrate this rattle to you.*
> *Thank you for infusing it with your powerful*
> *healing energy and light.*

I ask for and affirm that it will always serve
a true and positive purpose.
I ask for and affirm that in all ways and at all times
it will purify, liberate, and bless.
Thank you, thank you, thank you.
Blessed be. And so it is.

Now hold it to your heart for a moment to further align yourself with its energy.

Periodically, and whenever you'd like to refresh its potency, put the old garlic pieces at the base of a tree and cut a fresh garlic clove into eight pieces and place them in the shaker. To keep the energy of the rattle fresh, replace both the beans and the garlic every year or two by placing the old ones at the base of a tree and repeating the entire process (sage/sun, blessing, etc.).

Note: If you prefer not to use garlic because you're concerned about the scent, you can leave the garlic out or replace it with something more fragrant, such as eight whole pieces of allspice, anise, or cloves.

Using Your Rattle

Now that you're the proud owner of a powerful space-clearing tool, stand in a room of your home and move in a generally counterclockwise direction around the area while loudly shaking your rattle. Pay special attention to the corners, under tables, behind doors, and anywhere else you imagine energy might be likely to get stuck. As you shake, you might visualize—or just be aware of—old, stagnant energy being unstuck and consequently transmuted into bright, sparkly, positive energy. Repeat this process in every room and area of your home.

Important: Be mindful of animal companions during this process. The loud sound of the rattle can frighten them, so it's best if they're in a different room or outside.

Again, our objective right now is just to shift the energy in a positive way. Still, as I mentioned, the above rattling exercise—especially when it's coupled with the following practice—can often be all that's needed to get the energy moving in the right direction. At any rate, it's a good start! We'll be getting a bit more thorough later (for those more complex or long-standing energetic challenges), and your shiny new rattle will come into play then, too, so keep it handy!

PURIFY WITH SMOKE

Fumigation, or smudging, is another tried-and-true shamanic/magical technique for purifying the energy and vibration of a space. Later we'll learn about various smudging tools and techniques, but for our current purpose, simply obtain a bundle of dried white sage. (White sage bundles are available online and at many health and metaphysical stores. You can also pick, wrap, and dry your own; just make sure to honor the plant by asking permission first and offering a small libation of water, beer, or wine as payment and an expression of gratitude.) Hold the bundle of sage in your open, cupped palms, and say a sage blessing invocation such as this:

By north, south, east, and west
This sage is consecrated and blessed.
By ground below and stars above
This sage brings peace, purification, and love.
Mother Earth, I thank you.
Father Sky, I thank you.
Spirit of Sage, I thank you.

Now light the sage and shake to extinguish the flame so that it's smoking but not flaming. If necessary, light and extinguish it again until the smoke is flowing. Carrying a small dish to catch burning embers, move in a counterclockwise direction through the room, letting the smoke swirl around and douse the perimeter. Pay special attention to the corners and any dark or out-of-the-way areas. As you do this, visualize/imagine/notice the smoke clearing and fine-tuning the energy of the space and bringing it into a very positive and clear vibration. Repeat in each room and area. (Oh, yeah—and you might want to temporarily disable your smoke detector and then immediately re-enable it as soon as it's appropriate!)

When you're finished with the house, move the smoke around your own body to purify your personal energy field and fine-tune your vibration.

Extinguish the sage by sealing it in a jar or sticking it in a cup of sand.

To complete the smudging process by setting a new and more positive energetic tone, stand in a central location in your home. Close your eyes, and hold your hands in prayer pose. Visualize a sphere of very bright golden-white light (like the sun) completely filling and encompassing your home. Say a prayer or invocation to call in positive energy and blessings, such as this:

God/dess (or Universal Light), thank you for filling
this home with love, light, clarity, serenity, abundance,
creativity, laughter, and happiness. Harmony now fills it
and harmony surrounds it. Peace prevails now and always.

Great work: you have successfully shifted the energy in your home (and life) in a major way! I highly suggest that you now recalibrate or balance yourself by lying flat on the ground for a bit. Take some deep breaths, relax, and visualize any excess energy you may have generated or called into your energy field as falling downward into the earth like water dripping from freshly washed clothes on a clothesline. When you feel ready, stand up and have a bit of cereal, rice, or bread; this will help to further ground your energy.

Note: The dried sage bundle can be substituted with a mister containing water and twenty to thirty drops of essential oil of sage. Bless the mister as you would the sage bundle, then simply mist instead of smudging.

Choose a Talisman
(or Let a Talisman Choose You)

Whether you consciously sensed it or not, if you performed the above exercises, you successfully shifted the energy in your home for the better. If you think about it, you'll realize that you affected the subtle realm by taking steps in the physical realm (such as shaking a rattle and lighting a sage bundle). In other words, you began to experience firsthand what it means to walk between the worlds. And walking between the worlds (of form and spirit, matter and ether, seen and unseen) is what defines the work of shamanic healers, magical practitioners, and energy workers of all varieties because it is the point of power, or the place where everything unites and where we can consciously wield our spiritual power to affect the positive changes we desire.

And, because you'll be doing even more of this between-the-world walking in the chapters ahead, if you feel drawn to the idea, you might like to obtain a talisman, or an object or symbol that will help you to stay centered and grounded in yourself and to feel safe and protected no matter where your spiritual healing work may lead. So, naturally, if you do choose to obtain a talisman (at least at first), you'll want to keep it with you whenever you consciously venture between the worlds.

You may already have your talisman—it may merely await your recognition. Otherwise, finding it may require a walk in the woods or a trip to a rock shop or metaphysical bookstore. The important thing is that it "clicks," or resonates, for you as something that will be a powerful protector and companion in your travels between the worlds.

Here are some talisman options:

- A rock or crystal
- A piece of jewelry
- A rune
- A stick from a beloved or befriended tree
- A piece of driftwood
- A seashell
- A small statue or figurine
- A tattoo

If something doesn't click for you right away, don't force it. Simply keep an eye out for your talisman and be mindful of your intention to discover it. To help this process along, you might simply consider your desire to connect with your talisman just before you drift off to sleep. As you do so, feel the feelings you want to feel when you gaze at your perfect and divinely selected talisman, such as confidence, power, and inner knowing. This will begin to magnetize it into your life experience, and when you discover it, you'll know.

Talisman Consecration Ceremony

Once you've discovered or been united with your talisman, perform the following ceremony to consecrate it to its new purpose and help align you with its energy. Perform it on the night of a full moon, when the moon is high in the sky. (If your talisman is a tattoo, skip to the tattoo talisman consecration ceremony on page 28.)

If appropriate, begin by physically washing your talisman with water and perhaps sea salt and/or natural soap. Then dry it.

Spread a white cloth on an altar or tabletop, and place your talisman in the center. Encircle it with four tealight candles (soy or vegetable wax is always best for you and the environment) at the four cardinal points: north, south, east, and west. Then place four more tealights around it in between the first four, at the northwest, northeast, southwest, and southeast points. After you light the tealights, turn out all the electric lights.

Light your bundle of sage, and smudge your entire body to purify your personal vibration. Then carefully pick up your talisman and smudge it as well. Just as carefully, place it back within the circle of candles and extinguish the sage.

Visualize a sphere of very bright golden-white light encompassing and filling your talisman. Say:

> *North, south, east, and west*
> *This full moon night you're eight times blessed.*
> *By day and night and night and day*
> *Between the worlds you'll light my way.*

Continue to direct positive energy toward the talisman for as long as you can hold your focus, and accompany this with any additional prayers and blessings that feel right.

Then pick up your talisman, and hold it to your heart for a moment. Extinguish the candles. Keep your talisman with you or place it somewhere where it will not be disturbed, perhaps in a special box or bag that is expressly for the purpose of keeping it safe.

TATTOO TALISMAN CONSECRATION CEREMONY

No matter how long you've had a tattoo, if it feels right, you can put it to work as your personal talisman. (Just make sure it's 100 percent healed before you perform the following ceremony.)

On a full moon night, when the moon is high in the sky, take a shower or bath and then assemble your ingredients.

YOU WILL NEED:

> 4 tealight candles (ideally soy or vegetable wax)
>
> A bundle of white sage
>
> A small bottle or jar of jojoba oil into which you have added 10 drops of essential oil of myrrh if you're female or frankincense if you're male, or a combination of both if you prefer

Note: Make sure the oils you choose don't irritate your skin. If you have highly sensitive skin, forgo the essential oil completely. Instead, cleanse a small white quartz point with sage smoke and then place it in a small bottle or jar of jojoba oil or another carrier oil that you know won't irritate your skin.

Position the tealights around your body at the four cardinal points so that you have room to safely stand in the center. Light the candles and stand in the center, facing north. Light the sage and smudge yourself, paying special attention to your tattoo.

Visualize your tattoo glowing with very bright white light as you say:

North, south, east, and west
This full moon night you're eight times blessed.
By day and night and night and day
Between the worlds you'll light my way.

Anoint the entire surface area of the tattoo with the jojoba mixture. Then anoint your belly, heart, and third eye with a dab of the oil as well. When it feels right, extinguish the tealights (you can use them again at a later date if you choose). Then every night until the next full moon, dab a little of the oil on your tattoo before you go to bed.

An Etheric Talisman Option

For those of you who are especially comfortable with visualization, you might like to have an etheric talisman. In other words, before walking between the worlds—or anytime you feel you could use some extra protection and courage—you might visualize a specific symbol that feels powerful to you, such as a glowing royal blue, five-pointed star above your head or a bright, silvery-white crescent moon on your brow. If you feel drawn to this type of talisman, I suggest performing the following ceremony:

Etheric Talisman Consecration Ceremony

On a full moon night, take a bath and dress comfortably. Then, by candlelight, close your eyes, relax in a comfortable seated position with your spine straight, and take some deep breaths. When you feel centered and prepared, summon your visualized talisman. See it as clearly and as vividly as possible.

Then visualize and imagine it bathed in golden-white light as you say:

North, south, east, and west
This full moon night you're eight times blessed.
By day and night and night and day
Between the worlds you'll light my way.

A Note on Your Talisman

Be aware that the protective power of your talisman comes from your own intention, the spiritual energy the previous exercise helped infuse it with, and the mental energy you've directed toward it. All of these ingredients are available to you with or without the talisman, so there is no reason to become superstitious about your talisman or to feel that you are in any way unsafe without it. It is simply a tool that can help you to generate and to maintain the feelings of focus and confidence that are ideal for walking between the worlds, which is why you may find your talisman to be a trusty companion for many of the exercises you'll learn about in the following chapters.

three

Enlisting Help

he single best thing we can do to improve our personal energy and the energy of our home does not involve lighting a sage bundle. Neither does it involve shaking a rattle or even setting an intention. While all of these things are wonderful techniques that help shift the vibe in a wonderful way, the very best and most powerful thing we can do when it comes to drumming up the good vibes and positively affecting our life conditions is to ask for help.

The Divine Presence (aka God/dess, Universe, All That Is) is infinitely available to help us with every single aspect of our life experience. Provided our intentions are purely positive and for the truest good of all concerned, all we have to do is ask and allow perfect, endless, all-encompassing help from the Divine to come rushing to our assistance.

And so, while we'll get to more formal and comprehensive forms of energetic clearing, shielding, and fine-tuning in later chapters, it's important to begin with an orientation to the art of divine delegation. After all, any endeavor can be exponentially enhanced by an initial request for divine assistance, and often a simple prayer or invocation can be all that's needed to take care of the whole thing without us even having to lift a finger.

A Few Words on Personal Spirituality

"God is a metaphor for that which transcends all levels of intellectual thought. It's as simple as that."

"Every religion is true one way or another. It is true when understood metaphorically. But when it gets stuck in its own metaphors, interpreting them as facts, then you are in trouble."

"I don't have faith, I have experience."

—Joseph Campbell

I am a firm believer that all true spiritual experiences are 100 percent personal and 100 percent unique. When we begin to subscribe too rigidly to a particular canon of beliefs, we surrender our real, direct experience of the Divine. Furthermore, the Divine, being a "metaphor for that which transcends all levels of intellectual thought," necessarily transcends all rigidity and rules anyway. By definition, it cannot possibly be pigeonholed, much less reduced to any single interpretation or belief system.

Because it transcends intellectual thought, our spirituality, in essence, might be defined the same way as our creativity. So when an image, symbol, concept, story, or idea evokes the presence of the Divine within your consciousness—when it allows you to transcend intellectual thought and connect with that place that is between the worlds—it is truer than true and realer than real: it is the Divine speaking to you and you speaking to the Divine. And, just as no two artistic sensibilities are exactly alike, no two individuals can possibly have the same experience of the Divine.

With that in mind, you might think of the following descriptions as facets of the One Infinite Jewel, rays of the One Infinite Sun, or conceptual or finite doorways into the Inconceivable Infinite. As you read through them, be alert to which ones really resonate with you, and ignore the rest or file them away for later. Or ignore them all and work with whatever interpretation(s) or facet(s) of the Divine speak to you and give you the feeling that they will be powerful allies for your spiritual work.

A FEW WORDS ON REQUESTING HELP

What we're really doing when we request help might be described as accessing the part of our own consciousness that is one with the Divine. Of course, in truth, every part of us is one with the Divine, but because we're operating under the illusion of separation so much of the time, to access our divine power we must transcend our egos, and the best way to do that is to ask for help. Put another way, simply calling on the Divine—with any name and in any form that feels powerful for you—takes you beyond the "little me" and allows you access to infinite power as long as you (a) ask for something that is for the truest good of all (because the Divine is not in alignment with anything else), and (b) trust that your request has been heard and is being acted upon in the best of all possible ways.

To summarize:

- Ask for help.

- Ask for help from an aspect (or aspects) of the Divine that feels powerful for you (more guidance on this in the sections below).

- Ask for help in a way that feels powerful for you (whether this involves a prayer, an affirmation, a visualization, a ritual, or all of the above).

- Ask for something that is for the truest good of all concerned (which definitely would include clearing and shielding yourself, improving your personal vibration, and improving the vibration of your space).

- Trust that your request has been heard and is being answered in the best and most effective possible way.

PERSONAL ALLIES

We never walk alone. Wherever we go and whatever we do, spirit helpers and beings of light accompany us. Some of these—which we'll talk about in the next section—are universal: in other words, they simultaneously accompany many people in various parts of the world and at various times throughout history. Others are part of our own unique energy fields: they hang out with us personally and assist us with our spiritual work. You might think of them as aspects, or kaleidoscopic facets, of you that exist on other planes of reality. The aspect of you that exists on this plane, then, would be like a spiritual ally to them as well. The idea is that when you and one or more of your personal spiritual allies align your purposes and effort toward a specific aim (which is a way of walking between the worlds), you can be exponentially more effective.

Meeting one of your personal allies (or several of them) can be a very powerful experience. Once you make contact with an ally, you can receive important information and messages from him or her through your sixth sense, which may include (but may not be limited to) messages received via your thoughts, feelings, dreams, daydreams, inner vision, inner hearing, inner knowing, or meaningful, recurring experiences in the physical world. You can also ask your personal allies questions or for specific types of help, and then listen for or be alert to messages received via one or more of these sixth-sensory avenues.

Now I'll introduce you to the main types of personal spiritual helpers. Then, if you feel inspired, you can take the shamanic journey (visualization exercise) at the end of this section to meet one or more of your personal allies who are especially qualified to assist you in your energy clearing and spiritual healing work.

Animal Spirit Guides

Many spiritual healers and most shamanic practitioners have a relationship with one or more animal spirit guides. For example, while meditating once I had a very clear vision of a male lion. I felt a very distinct affinity for this lion, and I knew that the lion was an aspect of my personal energy field and the embodiment of my confidence and personal power. I also knew that the lion had a name, and that the lion's name was Rasa. Even though that meditation was over ten years ago, a deep awareness of Rasa's presence has stayed with me. I sometimes ask him for help, and I know that even when I don't ask, he's helping me with confidence and personal power–related issues all the time.

Some animal spirit guides may stay with you for life. Some may come and go at different times in your life for different purposes.

Guardian Angels

Author Doreen Virtue says that every human being has at least one personal guardian angel (and sometimes more) that stays with him or her from birth until death; I have an intuitive sense that this is true. I also believe—because I've personally experienced it—that when we ask for help from our angels and consciously include them in our daily lives, their light gets brighter, and they seem to be more powerful and present. This would make sense from a metaphysical standpoint: humans are characterized by free will, and so even if an angel is dedicated to helping us throughout a lifetime, the angel is unable to lend the tiniest bit of help unless we request and accept it.

Sometimes I can even detect the presence of guardian angels with my physical eyes: they appear as brief flashes or sparkles of white light next to the place where I'm standing or sitting, or near the top of someone else's head. Perhaps you've noticed something similar or will begin to now that you're alert to the possibility.

We can also consciously tune in to the presence of our guardian angels just by relaxing and asking them to reveal themselves to us. You might close your eyes and inwardly say, "Guardian angel or angels, please make yourselves known to me." Then see if you can sense one or more beings of light near your shoulders.

Once you sense this, you might ask their names. Then relax, and allow your mind to be receptive to anything you hear. The answer might appear as a thought or a flash. Then, in the future, you can call on your angel(s) by name and ask for help and protection whenever you desire it.

Because intuition comes through the same channel as imagination, when you first begin to tune in to your intuition in this way, it can sometimes feel like you're making things up. Instead of getting too serious about the process and discounting the information you receive, just make a note of whatever comes to you, and approach the whole thing with a sense of playfulness and fun. Over time, you'll become surer of your intuitive hits.

Spirit Guides

By "spirit guides," I mean personal allies who appear in human form rather than appearing as angels or animals. It's possible that they're humans who once lived on our plane of existence and have since crossed into the light. It's also possible that they're an archetypal manifestation of a certain aspect of our own personality or perhaps a facet of our own souls from a previous or future lifetime (or simultaneous lifetime if time is not linear, as I suspect it is not). Regardless, it appears that spirit guides are present because they have an affinity with us for some reason, such as our ancestry, a particular past life, or our unique spiritual mission in this life.

For example, one of my spirit guides is an old Chinese mystic with a long beard and a staff. He helps me with understanding the principles of feng shui and Taoist alchemy. Another one of my spirit guides is an old Native American woman who helps me to recognize the unique spiritual wisdom and healing uses of various types of plants. As I alluded to above, perhaps these guides are hanging around because our purposes are symbiotic or because they are archetypal aspects of my own personality. The mechanical explanation of their presence is shrouded in mystery, but this does not diminish the functionality of my relationships with them or the fact that they infuse my life experience with deeper levels of texture and dimension.

Spirit guides might stay with you for a lifetime, and some may come and go for different purposes and at different times.

Shamanic Journey to Meet Your Ally or Allies

This visualization ritual will help you meet a personal ally or allies who will be significantly helpful to you at this stage in your spiritual path.

In a darkened or soothingly lit room where you won't be disturbed, light a candle and play relaxing and entrancing music, such as the *Shamanic Dream* album by Anugama. You might also light a stick of incense that assists with visioning, such as Nag Champa, dragon's blood, kyphi, or copal.

Sit or lie down comfortably, with your spine straight. Relax, center yourself, and take some deep breaths. Consciously set the intention for this visualization to meet one or more personal allies who would like to assist you with staying energetically clear and positive. When you feel ready, close your eyes and take yourself through the following exercise; if you feel you might have trouble easily remembering the steps, you might record your voice or have a friend read this aloud to you:

Begin by reminding yourself that you are safe. You are free to relax completely, because as you take this journey you are divinely protected and guarded in every way. As you relax, no harm will come to your body, mind, or affairs, so you are free to let go of any and all worries and surrender completely to the journey.

Pay attention to your breath and allow it to slow down and go deeper. Relax more and more completely with each breath in and out.

Now consciously scan your body and relax any area where you may be holding on to tension. Remind yourself again that it is safe to let go. Breathe into any tension and allow it to dissolve, evaporate, and disappear.

Now that you're relaxed and feel completely safe and protected, find yourself—in your mind's eye—in a forest. Take a deep breath in through your nose and notice the scent of the trees and the earth. Look at the leaves on the trees and the texture of the bark. Listen to the sound of a light breeze through the leaves.

The temperature is perfect. Feel it on your skin. It's not too hot and it's not too cold. It feels completely nurturing and completely comfortable.

Look down. What is on your feet? Perhaps you're barefoot, or perhaps you're wearing boots or sandals. Once you've taken note of your footwear, move your attention to how soft and supportive the ground feels underneath you.

Now look at your body. Are you wearing any clothes? If so, what are you wearing? Whatever it is, realize how comfortable it is and how completely perfect you feel in it.

Suddenly you realize that while everything else is comfortable, there's one thing that isn't. It's a heavy backpack that you're carrying on your shoulders. You realize also that you don't need this pack. It represents everything that has been holding you back—everything that you no longer need and that never really defined you in the first place: your fears, insecurities, grudges, seeming physical and mental restrictions, and limiting beliefs.

With a great effort, you swing this pack off your back and hold it in front of you. Though you've carried this for a long time, you now make the decision that you no longer need it, and that it's time to discard it. You drop it like a ton of bricks.

Now, as you look down at it, you realize that it is decomposing very rapidly. It's breaking apart and decaying and becoming soil. The forest knows that you no longer need this pack, so it's dissolving it for you and transmuting it into compost. Before you know it, it's completely disappeared into the soil, and any negativity it held for you is transformed into rich, fertile soil.

You suddenly realize how free you feel! You are lighter, freer, and more joyful than you have ever felt in your life. Because you feel so joyful, you begin running though the forest. Running almost feels like flying, and each step you take upon the earth feels more and more empowering and enlivening.

You are filled with so much energy, you continue to run, and you find yourself running up a beautiful mountain path. Instead of becoming tired, you just keep feeling more and more energized and invigorated.

After a time, you realize that you've gone very high up the path, because you see that there's a beautiful view beneath you. This causes you to slow your stride to a brisk walk, but you still feel more and more filled with energy and enthusiasm as you ascend.

Drink in the gorgeous view all around you. Notice the forest floor below, the beautiful sky above, and the plants, rocks, and any bodies of water that may be present. Perhaps you also hear or see birds and other forms of wildlife. Let it all be a joyful symphony for your senses.

When you feel that you've reached the highest point of the mountain, you notice a very enticing gate or door in the side of a cliff, and you know that it has been put there expressly for your benefit. What does this gate or door look like? How big is it? What is it made of? How does it make you feel to gaze at it?

Finally, when you feel ready, you reach out to open the gate or door, and it swings open easily because it has only been awaiting your touch.

Before you even enter, an otherworldly light with a magical quality shines through. Then, as the opening gets larger, your eyes drink in a vision of more loveliness than you've ever encountered in your life. And although you've never seen this place before, you know it; it's as if it's a part of your very soul.

Slowly, reverently, you enter. Where are you? Look around. Soak in the sights. This is your magical place. Let yourself bask in its power.

Then you look up and see someone approaching. Who is it? It might be one being, and it might be more. As the being approaches, you feel a tremendous sense of love. This being's presence makes your heart swell with joy. Allow him/her/them to approach and then simply be with him/her/them. You might have a conversation out loud or engage in silent conversation via feelings and a sense of inner knowing. Or perhaps the being or beings have a gift or a message for you that will help you on your spiritual path and energy-clearing work. Be alert to the exchange, and allow it to unfold as it will.

When this feels complete, thank the being or beings for their presence and help in whatever way feels right to you. Then say goodbye, knowing that this being or beings will be with you in spirit and have been with you in spirit since before you even realized it.

Go back out the gate and find your way back down the mountain path. When you get to the bottom, envision yourself touching a tree trunk to begin grounding your energy.

Then, slowly, come back into an awareness of the room. Wiggle your fingers and toes; when you feel ready, open your eyes. Relax for at least three additional minutes.

Before you stand up, you might like to jot down some of what you experienced in your journal.

Finally, eat a little something grounding (such as root vegetables, grains, or legumes) to remind yourself and your body that you're back on the physical plane.

How Personal Allies Can Be Helpful for Spiritual Healers and Energy Workers

You might find that consciously aligning with a personal ally before or after you engage in energy work can help you stay focused, receive helpful intuitive messages, perceive the subtle realm, and clear, shield, and fine-tune more effectively. You can do this by calling on and/or visualizing your ally or allies before you perform a clearing and simply asking for their help. You might eventually begin to be aware of your ally by your side as you work, actually clearing right along with you or offering you moral support. If and when this happens, you can ask them to perform certain aspects of the work and/or ask them for helpful advice and information about the energetic and spiritual state of things in your immediate vicinity.

As you lie in bed at night before you fall asleep, you can also mentally ask for an ally to appear in your dreams or affect your dreams to help you sharpen your sixth sense and spiritual/magical skills. Know that this will almost definitely occur, whether you consciously remember it or not.

Place-Specific Allies

When you begin to tune in to the energy of a given place, you might discover (through your intuition) that there are local spiritual helpers who are ready and willing to help clear stagnant energy and liberate trapped entities. These types of beings are sometimes called psychopomps.

Fairies, Devas, and Nature Spirits

Every plant has a spirit and a divine wisdom and power. Additionally, natural settings are often populated by a number of local nature spirits. If it feels right or if the mood takes you, you might choose to call on one or more of these beings for help with your spiritual path and energy-clearing work.

Animals

In many forms of Native American spirituality, there are individual animals and then there is the one collective spirit, or divine representation, of any given species. If you're performing space clearing or any other energy work in a home that's located in an area that's native to an animal that feels powerful for you, you might choose to call on the spirit of that animal to help you in your work.

Many types of animals are famous for their ability to act as psychopomps. These include:

- Owl
- Cat
- Dog
- Coyote
- Butterfly
- Wolf
- Snake

But you need not limit yourself to this list.

Saints and Divinities

Oftentimes a particular saint or divinity is linked to an area or region. For example, Venice Beach, where I live, is named after Venice, Italy, which is named after the goddess Venus. There are, of course, no coincidences, and I do indeed feel the presence of the love goddess in this little enclave of the city of angels (another significant name). Right next to Venice is Santa Monica, which is named after Saint Monica and has a completely different vibration than Venice, despite the proximity of the two regions; Santa Monica is still feminine but much less wild and much more modest.

There are other indications a saint or divinity might be linked to a certain area, such as a historical association or the name of a street, lake, or other landmark. Or you might receive an intuitive hit that reveals a particular divinity or saint who happens to linger around the region. Any of these beings would be great choices to call on for help, as it's generally in their best interests for the energy of their stomping grounds to be as clear and vibrant as possible.

UNIVERSAL ALLIES FOR SPIRITUAL HEALING AND ENERGY CLEARING

And then, of course, there are those helpful beings that specialize in energy work, space clearing, and general positivity. There are many different divinities from many different cultures with these types of specialties, but here you'll find some that I've personally come to know and love.

Archangel Michael

I have found that Archangel Michael is the ultimate ally for all forms of clearing and energy work, including ushering earthbound entities into the light. His power is swift, potent, and all-encompassing. He carries a sword of light that can create a window between the worlds and cut any cords of attachment that an earthbound spirit (i.e., an unhappy ghost—more on these in chapter 5) may have to the physical world. He can also help lovingly escort these wayward spirits into the light.

What's more, I find him to be unsurpassed in his abilities to establish a very clear and positive vibration in the space (which naturally dissuades all forms of negativity) and to create a shield of protective light around people and spaces. To enlist his help with any of these aims, simply request it.

Saint Germain

Saint Germain, the divine bearded metaphysician, is quite an interesting fellow. He's an alchemical genius who specializes in something called the violet flame, which is a transmuting energy that changes negativity into positivity in exactly the way that is most needed. So, for example, you might call on him to transmute stuck or stagnant energy into positive energy, or to recalibrate the energy of a space after an argument or illness, or after an entity or entities have been cleared. Simply ask him to use his violet flame, and then visualize the space being filled with it, burning away and transmuting negativity into positivity and blessings.

Merlin

You may know Merlin as the archetypal alchemist and trickster spirit from the Arthurian legends or as the wise old magician who shows up again and again in the dreams, stories, and collective imagination of our culture. (For example, Gandalf and Dumbledore are two of his many incarnations.) He's an excellent ally to call on for help with deepening your skill and confidence when working with the subtle realm, and he can fine-tune your energy field and the energy field of your home like no one else. Once you make contact with him through meditation and visualization, and once you demonstrate the sincere desire to work with energy for the truest good of all, Merlin will be happy to assist you with any magical or shamanic endeavors you may embark upon. All you have to do is ask.

Rainbow Woman

Rainbow Woman, also known as Ixchel, is a Native American/Mayan deity who can establish a powerful doorway of light throughout the entire space. This doorway can gleam and glitter with rainbows, soothing and swirling away negativity and trapped energies. When you call on her before or during a space-clearing ritual, you might request her help by powerfully thinking or saying these words:

> *Rainbow Woman, I call on you! Please open up a large*
> *doorway of light throughout the entirety of this home!*
> *Please transmute negativity into positivity and*
> *liberate trapped energy by weaving a song*
> *of sweetness and glimmering color.*

White Buffalo Calf Woman

White Buffalo Calf Woman's purpose is to establish peace on earth, and she cares very much about aligning each and every square inch of the planet with its truest and most divine expression. So, naturally, she's more than happy to help transmute stagnant energy into vibrant energy with her loving vibration that feels like pure, cool, vast, blindingly white sunlight.

When calling on her, begin by holding the vision of peace on earth. Then request her help with the intention to create ripples of positive and vibrant energy throughout the planet.

Brighid

Brighid is a Celtic fire goddess who really puts her foot down when it comes to negativity and stagnation. To enlist her assistance during a spirit-clearing ritual, light a red candle, open all the windows and doors, and call on Brighid to fill the home with a very positive, bright reddish-orange, fiery light. This will quickly persuade all forms of negativity to hightail it out. Then call on Brighid once again to create a fiery energetic shield to preserve the positive vibration within the space. You might also hang a Brighid's cross—a special type of protective cross made out of sticks or reeds—on the door.

Mother Mary

When children are involved, Mother Mary can be a great help. If you have a child who has been feeling oversensitive or frightened, you can call on Mother Mary to watch over him or her and purify the space in your home in a way that will help the child to feel comfortable and safe.

Mother Mary can also be a great help when the earthbound spirits of children are involved. As you'll read about in chapter 5, children's spirits (or fragments thereof) can sometimes get trapped on the physical plane because of confusion, tragedy, or fear. Mother Mary's energy is very gentle and compassionate, and her presence can help them to feel safe about crossing over into the light. Mother Mary is happy to do this work. All you have to do is ask.

NAMES FOR THE ONE

Many people find that calling on the One Divine Presence for help is a key aspect of their spirituality and energy-healing practices. If this is the case for you, it's important to do this in a way that feels right.

Here's what I'm getting at: I find that many people are like me in that—while I believe that all is one and all is divine—I tend to shy away from the word *God*. I think Eckhart Tolle summarizes my feelings about this perfectly when he says in *The Power of Now*:

> The word *God* has become empty of meaning through thousands of years of misuse. I use it sometimes, but I do so sparingly. By misuse, I mean that people who have never even glimpsed at the realm of the sacred, the infinite vastness behind that word, use it with great conviction, as if they knew what they are talking about. Or they argue against it, as if they knew what it is that they are denying. This misuse gives rise to absurd beliefs, assertions, and egoic delusions, such as "*My* or *our* God is

the only true God, and *your* God is false," or Neitzche's famous statement, "God is dead."

The word *God* has become a closed concept. The moment the word is uttered, a mental image is created, no longer, perhaps, of an old man with a white beard, but still a mental representation of someone or something outside you...

Still, if the word *God* feels right to you and evokes a non-judgmental, non-limited, non-exclusive sense of all-powerful divinity, by all means, use it. But if you're like me and you need another, less overly misused word to describe and call upon the Divine (provided you haven't already settled on one that you're content with), consider using one or more of the following names instead—especially if you feel an expansive and whimsical sort of energetic charge as you read it:

- Goddess
- Great Goddess
- Mother/Father God
- Universe
- Great Spirit
- Spirit
- All That Is
- Infinite
- Source
- Source Energy
- Christ Consciousness
- Divine

- Divine Spirit

- Divine Presence

Personally, I like to use a number of these interchangeably to prevent conceptual stagnation.

A Few Words on Altars

Assembling an altar isn't necessary, but you might find it helpful. An altar can help focus your mind on your intention to request and allow help; it can also powerfully bring in the presence of any nonphysical helper(s) you may choose to invite. You can assemble an altar expressly for the purpose of keeping your energy and the energy of your home clear and positive, or you can assemble one to provide an ongoing focal point for your spiritual and metaphysical endeavors. (Or really for any intention that feels right.) It can be as temporary or as permanent as you wish.

If you feel drawn to assemble an altar, here are some simple guidelines:

- Obtain a statue, framed picture, or other representation of the helper or helpers you'd like to invoke. (You can get creative with this—for example, if your name for the Divine is "the Universe," you might frame a postcard picture of stars or a galaxy. Or, if you'd like to invoke the local nature spirits, you might choose a pinecone or a large rock from outside.)

- Place one or two candles near the representation on a shelf, small table, or other designated surface.

- You can stop here or you can add other items, such as an altar cloth (like a tablecloth), crystals, flowers, incense, fresh fruit, or other items that inspire a feeling of alignment with your chosen helper(s).

- If you keep your altar up indefinitely, be sure to keep it fresh. Compost perishables as necessary (do not eat or otherwise use them, as they are supposed to be offerings for the Divine), dust, straighten, etc.

- You can also add written prayers, words, or special symbols to represent exactly what you'd like to call in or manifest at any given time in your life.

- Keep the altar looking fresh, attractive, and free of dust. Replace the offerings as necessary (flowers, fruit, and so forth), and dispose of the old ones in a conscious and respectful way (for example, compost them or place them at the base of a tree). Also, light the candles at least once a week or so, and allow the altar to be a reminder for you to express your gratitude, reverence, and awe inwardly or aloud.

four

MAGICAL HYGIENE 101

ou are spiritually open and intuitive to an uncommon degree—otherwise, you definitely would not have made it this far into the book. And because people like us are so sensitive to invisible energetic conditions, it's especially important for us to clear and shield our personal energy on a regular basis. Similarly, spiritual sensitives like us have a lot of what might be called spiritual or magical power floating around in our energy field, and when we don't make a point of consciously wielding and directing this power, it can cause us to feel unsafe and unsettled, and—as a result—to unwittingly manifest situations and conditions that we aren't particularly crazy about.

I learned all this stuff the hard way. I used to be something of a psychic sponge for negativity and constantly felt like I was on an emotional roller coaster that took me from depressed to anxious to precariously elated. All of that changed when I discovered and formulated the exercises in this chapter and began practicing them daily. I call these practices magical hygiene, and, because of the amazingly positive experiences I've had with them, I'm very pleased to be able to share them with you.

You'll find that a regular magical hygiene practice will allow you to feel a very deep sense of serenity, safety, and stability that will nourish your physical and mental health like nothing else. But that's not all! Because all of existence is a sea of energy, keeping our energetic body clear and positive creates ripples of positivity throughout the entire unified field and consequently brings blessings to the world.

Clearing the Energetic Body

Your physical body has an energetic counterpart that we'll call the energetic body. To get an idea of the nature of your energetic body, you might begin by thinking of it as a sphere of light that occupies the same space as your physical body while also completely encompassing your physical body and extending outward from it in all directions. Within this sphere, there are places where specific types of energy gather into wheels of light (called chakras), as well as rivers and channels of light (called meridians). These chakras and meridians—and the entire energetic body— are inextricably linked with your physical body, and therefore they interact with and influence your physical body.

Just like we might get dust on our shoes or spaghetti sauce on our shirt, in the course of a normal day, we may inadvertently pick up negative energy from people, places, and situations, and this can be especially noticeable once we begin to become conscious of the subtle realm. But it's nothing to worry about! It's just a natural side effect of the illusion-based world in which we appear to presently live. And, like a daily shower, performing a regular self-clearing visualization as a part of your regular magical hygiene practice can keep your energy fresh and clear.

DAILY SELF-CLEARING VISUALIZATION

I like to perform this visualization almost daily or whenever I intuitively feel that I could use some extra clearing. Of course, if you need a bit of time to get in the habit, or if you accidently skip a few days here or there, there's no need to freak out! It's just a helpful tool. At first, though, so that you can experience its benefits firsthand, I suggest trying it every day for an entire week. Then you might like to begin to work it into your daily schedule at a pace that feels doable to you.

It might seem slightly time-consuming at first, but once you get the hang of it, it won't need to take any longer than five minutes or so.

Sit comfortably, with your spine straight. Take some deep breaths. Consciously connect with the Divine in a way that feels powerful to you. Request and visualize a divine vacuum tube of light that enters your energetic body and vacuums away all darkness, stagnation, and negativity.

When this feels complete, visualize/imagine/feel yourself growing roots of light from your legs and the base of your spine. See/feel these roots going deep into the earth, passing levels of rock, water, and dirt. Continue visualizing this until your roots reach the earth's core. Notice/sense that the earth's core is filled with very bright golden light. As your roots enter this light, you realize that it is actually a very dense magnet that draws your roots in and anchors your entire body to the core of the earth.

Then notice that the light from the core is moving up your roots. Continue to visualize this until the light reaches your body. Then it enters the area at the base of your spine, which is your root chakra, a glowing wheel of bright red light spinning horizontally. The earth light immediately transmutes any negativity from this chakra and allows the light to become clearer, brighter, and more vibrant.

The light then moves up to the area halfway between your pubic bone and your belly button. This is your sacral chakra, a glowing wheel of bright orange light spinning vertically. The earth light similarly transmutes any negativity from this chakra, allowing the light to become clearer, brighter, and more vibrant. This continues to happen with each of the following chakras:

CHAKRA	LOCATED	COLOR	SPINNING
Solar Plexus	Halfway between your belly button and the center of your sternum	Yellow	Vertically
Heart	Center of sternum	Green	Vertically
Throat	Middle of your neck	Blue	Vertically
Third Eye	About 1 cm above the area just between eyebrows	Indigo	Vertically
Crown	The very top of your head	White and/or purple	Horizontally

Now visualize the earth light merging with your light and continuing upward, as if you were a tree growing upward into the sky. Continue to reach up and up. Once the tops of your branches exit the earth's atmosphere, spread them into the great beyond. Envision this great beyond as filled with very bright clear light filled with rainbow sparkles and starlight. Draw this light downward, as a tree would draw in sunlight. See this light moving down the trunk until it reaches the crown of your head. As it enters the crown chakra, allow the crown chakra to be cleansed and activated with this cosmic light. Continue this process with each chakra in turn as before, only moving from the top down this time instead of from the bottom up. Allow the earth light and cosmic light to merge beautifully within your energetic body. Finally, request once more that the divine vacuum go through your energetic field, removing any and all lingering traces of negative or less than vibrant energy. Give thanks to the earth, cosmos, and Great Spirit.

Then finish with a shielding exercise (see page 60).

Feel free to adjust this exercise according to your needs and spiritual orientation, as long as you hit the following key points:

- Clear/vacuum your entire energy field
- Connect with the core of the earth
- Clear/activate each of the seven chakras with earth light
- Connect with the light of the cosmos
- Clear/activate each of the seven main chakras with earth light

- Clear/vacuum your entire energy field again
- Finish with a shielding exercise (see page 60)

Sea Salt Baths

If possible, to keep your energetic body shipshape, take a sea salt bath once a week or so. Simply dissolve ¾–1 cup of sea salt in a warm bath and soak for at least forty minutes. Be sure to keep lots of fresh drinking water handy to replenish your fluids. The cool thing is that this works on both the physical and the energetic levels: if your physical and energetic bodies were carpets, this would be like getting them steam cleaned.

Special Self-Clearing Cures

While you'll ideally be performing the above visualization exercise (or one like it) daily and taking sea salt baths every week or so, sometimes you might need something extra. For example, you might feel like you're carrying around someone else's problems or like you've picked up an especially heavy load of negativity from some situation or other. This may manifest in any of several ways, including (but not limited to) depression, fatigue, muddled thinking, excessive worry, obsessive thoughts, addictive tendencies, or long-standing blocks in any life areas. If you feel that you could use a little boost when it comes to clearing your energetic body, you might try one of the following special cures:

> **Aquamarine Cure:** Wear or carry an aquamarine crystal to help refresh and detoxify your energetic body. Be sure to cleanse it daily by running it under cold water for thirty seconds. *Especially good for:* self-esteem issues, depression, fatigue.

Garlic Flower Essence Cure: Place four drops of garlic flower essence under your tongue once in the morning and once in the evening. (Again, flower essences are the energetic vibration of a blossom preserved in water with brandy. You can find them online and at many health food stores.) *Especially good for:* a systemic feeling of being "unclean," giving away your power to other beings (physical or nonphysical), a weakened sense of self, poor boundary issues.

Peppermint Cure: Place ten to twenty drops of peppermint essential oil in a natural body wash and ten to twenty drops of peppermint essential oil in a natural unscented lotion. Shake both well to blend. Whenever you shower or bathe, use the body wash. Afterwards, use the lotion. *Especially good for:* low energy, feeling dragged down by others' negativity or negative situations, confusion or muddled thinking, excessive worry, fear. or other energetic blocks.

Shielding the Energetic Body

Would you take a shower and then walk straight out the door with no clothes on? On an average day, probably not! Similarly, after taking steps to clear your energetic body, it's important to shield yourself. This will help preserve your positive vibration while helping deflect all forms of negativity.

Daily Self-Shielding Visualization

After you perform the above clearing visualization or one like it, perform a shielding visualization like this one:

Relax and take some deep breaths. Connect with the Divine in a way that feels powerful for you. Now ask the Divine (or the part of you that is one with everything) to surround you—your entire physical and energetic body—with very bright white light. See this light in your mind's eye as a blindingly bright sphere like the sun. Then ask that you be surrounded with very bright indigo-violet light. Again, see this light surrounding you like an indigo sun. Ask that this shield of light stay with you, and mentally infuse it with this intention: "Within this shield of light only love remains, and through this shield of light only love may enter." Know that this sphere of light will not prevent you from seeing the truth of things, because the only real truth is love. You will still be able to observe the illusion of negativity in the forms of challenging situations and emotions, but the difference will be that you will not absorb this negativity into your energetic body. This will allow you to be more effective in all ways and, in the words of Gandhi, to "be the change you wish to see in the world."

Special Self-Shielding Amulets

While the above visualization will shield you quite powerfully for twelve to twenty-four hours, there are times when you might find yourself in especially negative situations or places, or you might feel especially energetically vulnerable now and then for whatever reason. When this is the case, for a little extra juice, carry or wear one of the following shielding amulets:

Mirrored Pendant Amulet: Make or obtain a necklace with a mirrored charm. This mirror can be any shape, as long as you like it enough to wear it on a regular basis. It's best if it hangs just below your throat, with the mirrored side facing out (though it's fine if both sides are mirrored). Before you put it on, hold it in bright sunlight (being careful not to start any fires), and say a quick prayer or invocation to charge the pendant with your intention. You might say something like this: "Great Spirit, it is my intention that this pendant reflect all negativity back to its source so that I may be safe and protected in all ways. Thank you for infusing it with this intention. Blessed be. And so it is." *Especially good for:* deflecting all forms of negativity.

White Quartz Amulet: Wear or carry a white quartz crystal point, making sure to cleanse it with white sage smoke or running water once a day. Before you wear it and at least once a week or so, hold it in your right hand and mentally charge it with the intention to preserve your positive vibration and transmute negativity into positivity. *Especially good for:* general protection and helping change negative vibes into positive ones.

Garlic Amulet: Carry a single clove of garlic. I like to tie it in fabric and safety-pin it to the middle of my bra, but you can put it in your pocket, place it in a small pouch around your neck, or carry it in whatever way works best for you. But first, hold it in very bright sunlight or just over a candle flame and mentally charge it with your intention to keep all forms of negativity at bay. *Especially good for:* traversing spiritually challenging places such as bars, nightclubs, or situations and locations characterized by dishonesty, exploitation, and greed (for example, I find it's always a good idea to carry empowered garlic when I go to Hollywood).

Special Shielding Visualizations

And then there will be those situations when you suddenly find you need an extra protection boost, but it's too late for an amulet: an unforeseen yet somehow necessary jaunt through a dark alley, a cocktail party that just happens to be in a haunted house, or a sudden and inexplicable feeling of unease. If this happens, don't panic! (*Never* panic. Always remember that you are divinely protected and possess ultimate jurisdiction over your energetic domain.) Just perform one of the following special visualizations:

Indigo Refresh Visualization: Very simply, connect with the Divine and visualize indigo light being refreshed around you. See it becoming brighter and more potent. *Especially good for:* transmuting/purifying general negativity.

Mirrored Sphere Visualization: Visualize a sphere of mirrors completely surrounding you, keeping you perfectly safe and protected, and reflecting all negativity right back to its source. *Especially good for:* deflecting danger, ill will, and ill intentions.

Angelic Bodyguard Visualization: Call on angels, and visualize that you are surrounded by at least four tall, powerful, bright beings of light that will immediately protect you from any negativity or negative intentions that might be sent your way. *Especially good for:* infusing you with confidence and shielding you from harm.

Earth Alignment Visualization: Remember your roots of light from the clearing visualization earlier? Connect with these roots and establish a connection to the glowing, golden light at the core of the earth. See this light immediately traveling up the roots and filling your entire body. *Especially good for:* grounding, calming, and aligning you with the ancient and all-knowing wisdom of the earth.

MEDITATION MADE EASY

So, starting now, you'll be practicing the cleansing and shielding visualizations daily (or at least actively working up to practicing them daily). Right? Right. So guess what? You'll be meditating! In the past, you might have said you can't seem to meditate, because you picked up the idea that meditating means you have to engage in some special, super-esoteric chanting regimen or, even more impossibly, to "stop your mind from thinking." In reality, we can reap all the benefits of a daily meditation practice when we simply take a few minutes to relax, sit with our spines straight, and focus our attention on something—all of which are important aspects of the daily cleansing and shielding visualizations. You might call it spiritual multitasking.

Later, once you get the hang of the visualizations and you can finish them both in a matter of five to ten minutes or so, you can naturally allow your meditation practice to grow and evolve according to what feels right to you. But for now, the important thing is to get in the habit. Experiment and find a time that works for you, such as the morning, during your lunch break, or just before you hit the sack.

Here are just a few more words on getting in the habit:

Eat, Drink, and Be Positive

When it comes to magical hygiene, what you put into your body—and how you relate to what you put into your body—is quite important. As we've seen, your physical body exists within and constantly interacts with your energetic body. Everything—even your physical body—is made of energy. So if you imagine that your energetic body is a lake, everything that goes into it affects the clarity and vibrancy of the entire lake. Below are some guidelines to help you eat and drink in ways that nourish and clarify.

BLESS

It's good to get in the habit of blessing your food and beverages before you consume them. To do this, before eating or drinking, simply take a moment to relax. Then tune in to the Divine aspect of yourself or call on the Divine in a way that feels powerful for you, give thanks for your food or beverage, and request or visualize the food or beverage being filled with positive and vibrant vibrations. You might visualize it as being filled with bright white light.

DRINK LOTS OF WATER

Drinking at least half your body weight in ounces per day is a great way to help clear out both physical and energetic toxins and to refresh your energetic body. If you're like me, you might find it helpful to invest in a water bottle that you love. Somehow, this seems to help me keep up the habit.

And do your best to remember to bless your water before you drink it! You might hold it in both hands and say a prayer such as this:

> *Great Goddess, please infuse this water*
> *with vibrations of purification and love.*

Additionally or instead, you might visualize very bright white light coming down from above, entering the crown of your head, going down to your heart, through your arms, and out the palms of your hands and into the bottle. If you want to get fancy with it, you can change the color of the light according to your intentions for the day. For example, you might choose:

- Pink for romance and friendship
- Green for physical healing, heart expansion, and prosperity
- Blue for creativity and self-expression

EAT POSITIVELY VIBRATING FOODS

Tune in to what feels right for you, but generally speaking, certain types of food have less positive vibrations than others. For example, the foods listed below are usually brimming with positive, sparkly life-force energy, especially when they're organically grown:

- Fresh fruit, dried (unsulphured/unsweetened) fruit, and fresh fruit juice
- Fresh vegetables and fresh vegetable juice
- Nuts, seeds, and legumes (especially sprouted)
- Whole, unprocessed grains such as oats, quinoa, and buckwheat (especially sprouted)
- Herbal tea

On the other hand, the foods listed below are often low in life-force energy and are characterized by less positive and life-affirming vibrations:

- Meat (especially nonorganic/non–free range)
- Dairy (especially nonorganic)
- Eggs (especially nonorganic/non–free range)
- Gelatin
- Processed sugar
- Processed flour
- Nonorganic wheat (the integrity of its nutrition has been compromised by exploitative farming practices)
- Gluten (a protein found in wheat and a few other grains—for many sensitive people, it's too energetically heavy and sticky)
- Artificial colors and flavors
- Preservatives
- Anything grown in a way that exploits people or regions, such as non–fair trade coffee or chocolate

five

EARTHBOUND ENTITIES 101

et me remind you once again that—because you are always divinely protected, and because you have ultimate authority over your energetic domain—you have absolutely nothing to fear when it comes to the subtle realm. I want to start with this because I know the topic of earthbound entities, or unhappy ghosts, can be a source of chronic heebie-jeebies for some of us. But I feel it's important to address this topic because, as I have observed firsthand in my work as an intuitive counselor and feng shui consultant, earthbound entities can sometimes hang around people or spaces, leaving less than ideal energetic conditions in their wake.

The good news is, because we do have jurisdiction over our own space and energy field, we can often easily help clear them out and help them cross into the light. But just in case you're still a little unsettled by the topic of this chapter, please let me assure you that:

- There is no need to look for or consciously notice entities in order to effectively clear them from your home or energy field. All the exercises in this book can be effective for these purposes whether you consciously tune in to the presence of earthbound entities or not.

- Simply feeling positive and confident can repel the negative effects of (and presence of) challenging entity situations.

- Simply calling on divine help can often be all that's necessary for escorting earthbound entities into the light.

- If you already think you've got an entity situation on your hands and you worry that it's a little too serious to deal with on your own, you don't have to! You can always call in a reputable expert and let her or him handle it.

Now, all of that being said, when I say "earthbound entities," what exactly am I talking about? As you'll soon learn, I don't exactly know (nor does anyone). Still, in every culture and since time immemorial, humans have been reportedly encountering them. And in this chapter I'll do my best to remove as much of the "unknown" factor as possible so that you can feel that much more confident and knowledgeable about this aspect of the unseen realm.

The Ego

In order to discuss earthbound entities, I must begin by discussing the human ego. I've heard the ego defined as "the illusion of separation," and I find this to be a very useful definition. To elaborate, when we think that we are in some way separate from everything else in the universe—i.e., the Infinite (aka God/Goddess/All That Is)—we're experiencing an illusion, and that illusion might be called the ego.

Of course, quite often we all experience this illusion, and in many ways it even sets the tone for what we call this plane of existence: what the Toltecs call the "common dream," or the agreed-upon human condition. Within this common dream, the ego can serve some very useful purposes. It can help us get places on time, remember facts like our names and where we live, and otherwise navigate through the practicalities of what we call everyday reality. If we always identified totally with our oneness with the Infinite, we would forget our names and addresses, and time would lose all meaning for us. This wouldn't be bad, but it simply wouldn't lend itself to life as we know it.

However, when we mistake our ego for who we *really* are—one with the Infinite—we begin to run into challenges of every sort. And the fact is, this happens to all of us on a fairly regular basis. For example, anytime we feel slighted, take things personally, worry that we don't "measure up," or become fixated on things like the idea that our partner "doesn't love us enough," we are mistaking our egos for who we really are. This is, of course, nothing to be ashamed of; it's just the way it often seems to go. I suspect that even the most enlightened among us are not entirely immune. And, on the bright side, it's often the catalyst for the exact drama that has the potential to teach us precisely what we most need to learn at any given time.

Now that you've been introduced to this way of defining the ego, consider death. If we believe the illusion that we are separate from All That Is, death is a very disquieting prospect. It takes everything the ego knows and values and sends it off into God knows where to fight, face, or try to conquer God knows what. Or perhaps it's the end of the ego, and consequently the end of everything that exists—and certainly the end of everything that has any meaning or worth—to the ego. And because we all eventually transition from this plane of existence, this casts a terribly unsettling light not just on death but on all of life as the ego knows it.

And so the more the ego runs the show, the more fear we experience—not just about death but also about everything else. The more we remind ourselves of the truth of who we really are, the more love and trust we experience. This is probably why Jesus said that fear is the opposite of love. Love is based on truth and acceptance of the mystery, and fear is based on illusion and a deep-seated mistrust of the unknown.

Even though the ego is an illusion, the emotions and attitudes it fosters can build up a tremendous energetic force. This is why we can sense tension when we walk into a room where an argument has just occurred or why we instinctively put our guard up when we encounter a particularly negative person.

THE EGO'S ROLE IN THE PRESENCE OF EARTHBOUND ENTITIES

Mary Ann Winkowski, the author of *When Ghosts Speak*, has been able to see and converse with the earthbound variety of ghosts since she was a child. Her Italian grandmother used to take her to funerals so that she could help the deceased cross over into the light. Like shamans and folk healers from many traditions, she says that after people die, they often stay with their bodies for about seventy-two hours. After that, they either cross into the light—or they don't. According to Mary Ann, if they choose not to cross over for whatever reason, the light then disappears or somehow becomes unavailable to them, and they seem to find themselves, at least for a while, trapped, or "earthbound."

While there are a number of reasons why it's said someone might not cross into the light, they all have to do with ego, or the illusion of separation. These reasons include things like:

Confusion: A child might think that he is supposed to stay with his still-living parents or he'll get in trouble.

Fear: A man might have been brought up in a "fire and brimstone" church and have the idea that if he crosses over, he will be burned in eternal fire.

Attachment: A woman might have defined herself by her diamond necklace to the point where she feels it is impossible to leave it behind.

Revenge: A man might have been murdered and feel that he can't cross over until he has sufficiently terrorized his murderer.

Justice: A woman might feel that she must stay earthbound until she exposes the foul play surrounding her will's execution.

Heartbreak: A mother might feel so attached to her feeling of responsibility for her still-living children that she feels guilty about crossing over into the light.

Addiction: A man might have been so addicted to heroin that he wants to hang around living heroin addicts to feed off their feelings of being high.

To many energy workers and healers (myself included), it appears that "the light" is the portal to an awareness of the Infinite and of one's own infinite nature. Indeed, many people who have had near-death experiences describe crossing into the light as a feeling of being one with everything, where there is no time and where they are reunited with their deceased loved ones, who are described as being in some way a part of the light. The highly influential spiritual author Denise Linn had a near-death experience after being shot by a stranger when she was seventeen. She describes a blissful encounter with the Divine Presence, during which she saw and knew herself to be one with the eternal, timeless nature of the universe and everyone and everything in the universe. She credits this with setting her on the path of teaching others about the infinite nature of the universe and of our souls.

One can see, then, why the ego might be afraid of the light. The light represents the Infinite, and once one is consciously united with the Infinite, the ego is completely divested of its power. In other words, in the light of truth, the ego (which is an illusion) becomes a nonentity, which is its biggest fear. If someone's ego holds too much sway, it might make a last-ditch effort at self-preservation by digging in its heels during the transition between this realm and the realm of the Infinite.

EARTHBOUND DYNAMICS
(What Exactly Are Earthbound Entities?)

I don't claim to categorically understand the dynamics of earthbound-spirit phenomena (and am frankly suspicious of anyone who does). But based on the energetic patterns I've perceived, I do have a number of theories. Whether or not any of these are literally true, the concepts are helpful when learning to sense the subtle realm and also can help with your energy work in general.

Conventional Wisdom: The Trapped-Soul Hypothesis

While there is a diverse collection of hypotheses about earthbound-entity phenomena, it appears that a large number of people believe that a spirit simply gets trapped. According to this assumption, as a result of an ego complication (as mentioned above), no aspect of the person's spirit or soul crosses over or goes anywhere and is essentially in limbo until he or she finds the way to the light or someone intervenes to help the spirit find the way to the light.

If we are all one with the Infinite, and the ego is simply an illusion, it's a bit easier for me to imagine that what we call the earthbound entity is just a fragment of the spirit of the being that was once alive, or perhaps that the earthbound entity is some sort of etheric memory or imprint in the energetic matrix that we call earth. Both of these—at various times or simultaneously—might be the case. (More on these hypotheses below.)

The Soul-Fragment Hypothesis

Have you heard of the shamanic healing technique called soul retrieval? During a soul-retrieval session, a shaman or healer utilizes any number of practices (trance, drums, rattles, etc.) to help retrieve fragments of a person's soul that may have been left behind during this or other lifetimes. According to practitioners of this modality, when we feel some sort of strong negative emotion (shock, grief, loss, etc.), a part of our soul might splinter off and become lodged somewhere in time or space. When these fragments are returned, the stagnant energy that has been fragmented and trapped becomes reunited with us and is no longer stagnant. In its newly energized state, it adds to our vibrancy and personal power. And, simultaneously, we are able to heal the trauma that we felt during that long-ago time, which allows us to move forward in our present life as a more confident, empowered, and generally happy person.

According to the soul-fragment hypothesis, what we call earthbound entities might be particularly substantial fragments of a deceased person's soul—fragments that pack such a potent energetic punch that they become noticeable or in some way troublesome to the living. In other words, while a portion of this person's life-force energy has moved on, a very large portion has been caught in an ego trap and has consequently become stuck on the physical plane without a physical body.

Another Way to Frame the Soul-Fragment Hypothesis

Astral travelers like author Robert Bruce often describe us as having not just one body but several: an astral body, a physical/etheric body, a projectable double, and so on. Another way to frame the soul-fragment theory is that the earthbound entity is the etheric portion of the physical body of the deceased now that the physical body itself has perished. After all, the ego is blinded by the physical realm and believes that no other realm exists. Therefore, it seems plausible that it might in some way splinter off and believe itself to be the whole of the once-living organism's consciousness, when in fact it is (or was) only a part. In this case, however, I imagine that it initially would need to borrow a substantial portion of the spirit's genuine life-force energy in order to keep up the illusion of separation that is its signature.

Author and professional ghost hunter Katherine Ramsland alludes to this idea in her book *Ghost* when she muses that a ghost (an earthbound entity) might be like a phantom limb. The body is gone, the animating force is gone, but some lingering trace of consciousness or energy continues on as if it is still very much *not* gone.

The Memory- or Energetic-Imprint Hypothesis

In Los Angeles, where I live, it is always easy for me to find a freeway entrance, even if I'm not at all familiar with the area. I don't think this is just because there are a lot of freeways in Los Angeles. I think it's because you can sense the knowledge hovering around in the ether. Wherever my car happens to be, millions of other people have been in the same exact place before me and maneuvered their cars like clockwork to the freeway entrance. Provided I let go and trust, my decisions about where to go to find the freeway entrance will generally be correct.

But we all know that the energy from thoughts, actions, feelings, and events can hover around in the ether long after they initially occur. Perhaps this is why realtors are legally required to tell potential buyers that someone has died in a home and why even the least metaphysically minded among us would probably not want to live in a home in which a murder had taken place, no matter how beautiful or how reasonably priced the place may be.

And this is the premise of the memory or energetic-imprint hypothesis. A very extreme negative occurrence or emotion manifests itself in a certain area—such as a violent murder or acute, prolonged grief—and it in some way imprints itself on the area, perhaps at the moment that a certain person or people die. Then, when someone senses this negative energetic condition, or earthbound entity, they lend it attention (and therefore energy—especially if their reaction is negative), and it begins to resonate more powerfully as a result.

The Parasitic- or Masquerading- Entity Hypothesis

Imagine this: an entity, or the energetic charge left over from the ego of a once-living person, is wandering about the physical plane confused and disoriented. Perhaps this entity does not fully comprehend that it is disembodied, sort of like the way we can wander through a dream without any cognition that we are dreaming.

This entity then finds itself magnetically drawn to a particular place where there is a lot of energy on which to feed, such as the house of a well-loved or well-connected recently deceased person. This person's relatives and associates might be going through drawers and closets, getting ready to sell the house and otherwise tying up loose ends.

Suddenly the entity feels very powerful. The entity can embody the energetic imprint of the deceased resident while simultaneously drawing excess energy from the physical and etheric bodies of the living people that are hanging around. The more energy this entity can siphon from the situation, the more this entity can affect things in the home, such as the lights, plumbing, and locations of small items. This, in turn, scares the living, which produces even more of an energetic charge on which the entity can feed. Of course, the people hanging around will naturally yet erroneously assume that the entity is the ghost of their recently deceased acquaintance.

Conditions such as this might also be the case in some famous and often-visited locations such as haunted hotels and the now-museumlike manors of deceased celebrities.

A Word on Light Beings and Spirit Helpers

And then there are what we might call light beings and spirit helpers, which are not earthbound entities. These include the spirit guides discussed in the previous chapter as well as deceased loved ones and beneficent beings who have crossed into the light.

Earthbound means exactly that: earthbound. Light beings and spirit helpers, on the other hand, are in no way stuck or bound to the earth plane. They might make an appearance in your dreams, feelings, or imagination now and then, and send you little presents or words of wisdom from beyond, but then they appear to be able to cross right back into the light, according to their whims.

If you happen to sense the presence of spirits, you will easily be able to tell the difference between those that have transitioned to the light and those that are stuck on the earth plane, because the former confer a feeling of joy and loving support, while the latter confer distinctly negative feelings. (I will outline these feelings at length below.) Spirits that have crossed over into the light also have much more depth and have the feeling of being complete, well-rounded personalities rather than the somewhat two-dimensional personalities demonstrated by earthbound spirits.

I bring this up because once you become attuned to the subtle realm, you will almost definitely encounter light beings and spirit helpers (if you haven't already), and I don't want you to think I'm suggesting that you need to clear them out.

How Does Reincarnation Fit into All This?

You may or may not believe in reincarnation. I, for one, am firmly convinced of its verity. (If you aren't or aren't sure, I highly recommend that you read *Old Souls* by Tom Shroder, after which I'm almost certain you'll be a believer or at the very least be open to the possibility.)

But in answer to the question of how reincarnation fits into all this, I have to begin by paying my respects to the mystery of it all by saying *I don't know*. Still, as you may have guessed, I have a working theory. Here it goes:

We are all one. We are not just one with each other, we are one with all of life and with everything that is—in this world and in the entire universe. Consciousness, therefore, may be like the ocean, while we as individual humans may be like raindrops or waves: temporarily distinct while still a part of the vast ocean.

Additionally, time as we know it is an illusion. (In fact, Albert Einstein himself said, "Reality is merely an illusion, albeit a very persistent one.") Rather than things occurring as if in a timeline that endlessly moves in one direction, everything that we perceive is happening, has happened, or ever will happen is actually occurring at once, and each moment is like a single wooden horse moving around a huge carousel. The actuality of this, of course, would be impossible to describe or fully conceive of with our human brains that are so accustomed to our little concept of linear time and with our language that relies on time and is, in fact, composed and absorbed in a thoroughly linear way.

With that in mind, many conditions or scenarios might be possible, including:

- Outside this plane of existence (like "the light"), the constraints of time are different or nonexistent, and spirits can in some way dwell there and in the world simultaneously.

- As in the soul-fragment theory, a large part of the spirit might be able to move on into the light and be reborn, while some energetic qualities, or fragments, remain trapped in our present time and space. (Using the ocean metaphor, this would be like if the consciousness of a person was a wave and the earthbound entity was a bucket of water from the wave that was left to stagnate for a while before reuniting with the ocean.)

- Spirits are not required to reincarnate right away and can be earthbound or exist as spirit helpers for a certain period of time before they move on.

- Some spirits, such as spirit helpers and light beings, may choose to stay in the light rather than be reborn on the earth plane.

DETECTING EARTHBOUND SPIRITS

I want to reiterate that (unless you have specific reasons for wanting to do so) there is absolutely no need for you to go out of your way to detect the presence of earthbound entities. Still, since you're going to be walking between the worlds on a regular basis, it will be helpful for you to be aware of some of the indications that they might be hanging around. Furthermore, becoming aware of their qualities and dynamics will help to dissolve any lingering fear so that you can traverse the subtle realm with confidence and courage.

Everyone's sixth sense is different, so you may intuitively detect the presence of earthbound spirits in a number of ways, including (but not limited to):

- An indescribable, otherworldly quality of light
- A level of quiet or a quality of sound that seems somehow notable or odd

- A temperature variation
- A sudden surge of a particular emotion that doesn't seem to be coming from you
- A singular odor, such as pipe tobacco or mustiness
- A slight taste in your mouth or a feeling in your throat
- Disembodied noises like walking or knocking
- An inexplicable knowing that an entity is near
- An inner vision or knowing regarding something that happened in your home
- A craving for something you don't normally crave (such as cigarettes)
- Dreams that point to the presence of entities

Sometimes you may also be able to detect or confirm the presence of earthbound entities through indications in the physical realm, such as:

- Chronic digestion issues. (Our stomachs are very sensitive to challenging conditions in the unseen realm, and digestion issues can sometimes indicate that our sixth sense is trying to tell us something.)
- Flickering lights or ongoing plumbing challenges. (According to renowned author and medium James Van Praagh, both these conditions can sometimes indicate the presence of earthbound spirits in the home.)

- Ongoing household discord. (Earthbound entities can feed off negative energy, and they have been said to incite unrest in the home for this reason.)

- One or more household members suddenly behaving in ways that seem to be uncharacteristic. (The energy of an entity can sometimes get intertwined with the energy of a living being, causing personality changes and challenges.)

- Children with imaginary friends. (Although in some cases imaginary friends are undoubtedly truly imaginary, it's said that since children are often highly psychically attuned, they can sometimes easily see and interact with earthbound spirits.)

Why We Want to Clear Earthbound Entities

Unless you are like my friend, the author Annie Wilder, and your life purpose is specifically related to hanging out with and supporting earthbound entities during this stage of their journey, it's usually ideal to clear earthbound entities from your home. (You will know this is part of your life purpose if you feel energized and inspired by the idea and presence of earthbound entities rather than muddled or drained.) This is because, generally speaking, an earthbound entity is stuck.

He, she, or it is a pattern of energy that has been held back from flowing in its most ideal and vibrant way, i.e., going into the light. It is temporarily disconnected from the Infinite and from the natural flow of things, and the aspects of its life that were once meaningful and satisfying are irretrievably gone. It is a stale and lonely bucket of seawater that has been separated from the sea. This means that if we are able to liberate it by helping it cross into the light, we are doing it a wonderful service while simultaneously making the energy in the area (and in the world) healthier, more vital, and more effervescent.

Like Attracts Like

Like attracts like. And since earthbound entities are generally characterized by stagnation and heaviness, simply cultivating flowing, sparkling, buoyant emotional and physical conditions can go a long way toward *un*inviting them and dissuading them from coming back. Furthermore, since the unseen realm reflects and interacts with the seen realm, and vice versa, having good boundaries in our day-to-day life fortifies our invisible boundaries and consequently prevents and discourages the unwelcome presence of earthbound entities in our home and personal energy field (see tables 5-1, 5-2, and 5-3).

Quick Note

In most instances, reading this book and engaging in the exercises it suggests will be all that you need in order to establish positive energetic conditions in your home and life. But if extreme conditions are present—such as long-standing depression or another form of debilitating mental illness—and if you suspect earthbound entities may have something to do with it, in addition to seeking immediate psychological assistance, you may want to call in a reputable expert in the field of entity clearing.

TABLE 5-1: **Conditions That Attract or Maintain Earthbound Entities**

EMOTIONAL CONDITIONS	Suppression
	Fear
	Anger
	Greed
	Sadness
	Mistrust
	Dishonesty
	Worry
	Depression
	Oppression
	Addiction
	Lack of consciousness
	Overemphasis on the external appearances of wealth or success
	Overemphasis on the external appearances of poverty or financial struggles
PHYSICAL CONDITIONS	Excess clutter
	Dirt and grime
	Stagnation
	Stale air
	Darkness and shadows
	Negative or depressing imagery
	Old items that hold negative energy
	Unhealthy plants
	Stagnant water

TABLE 5-2: Conditions That Repel or Prevent Earthbound Entities

EMOTIONAL CONDITIONS	The full range of emotions Honesty Happiness Joy Confidence Laughter Freedom Creative expression Consciousness Expansiveness Positive expectations Trust
PHYSICAL CONDITIONS	Minimal clutter Cleanliness Vibrant energy and movement Fresh air Light Bright and positive imagery Items that are blessed and that hold positive energy Vibrant plants Clear and free-flowing water

TABLE 5-3: **Activities That Attract or Repel Earthbound Entities**

ACTIVITIES THAT ATTRACT	Abusing drugs
	Addictive behaviors
	Playing with a Ouija board (this is because you are in effect sending out an open invitation to earthbound spirits to enter your personal space and energetic field)
	Pointedly calling in the presence of earthbound spirits, as you might during a séance (unless you are an expert)
	Engaging in shamanic rituals or energy work without regularly practicing good magical hygiene (see chapter 4)
	Engaging in shamanic rituals or energy work with the purpose of harming anyone or anything
	Sexual abuse
	Compromising yourself sexually in any way (doing things that don't feel right to you on a deep level)
	Compromising your boundaries in any other way, i.e., allowing things to happen that don't feel right to you or ignoring your instincts
	Physical or emotional abuse
ACTIVITIES THAT REPEL	Energy-clearing techniques
	Laughing
	Dancing
	Positive and uplifting music
	Creative endeavors
	Making choices that come from a place of self-love and self-acceptance
	Doing anything that makes you feel happy and energized
	Yoga
	Meditating
	Chanting

six

SPACE CLEARING

*i*n our modern world, sometimes we get the impression that there is functionality on the one hand and beauty on the other, or that there is practicality and then there is whimsy, and never the twain shall meet. Of course, we know that all of this is illusion. In truth, nothing is mundane: everything is magical.

Space clearing—an ancient practice present in countless cultures, though the term was coined by the author and healer Denise Linn—is a powerful method of dispersing and transmuting stagnation and heaviness and establishing beautiful, buoyant, sparkling energetic conditions in any given physical space. It's one of my favorite magical practices because it instantly unites these seemingly disparate aspects of existence (seen/unseen, physical/spiritual) and reminds me that nothing is more magical than my everyday reality, and that my home is not merely a building but an enchanted retreat. It infuses my home with a timeless, otherworldly quality, clears away the illusions of discord, and anchors me in the beauty of the eternal now.

And since you've now been initiated into the subtle realm, made the initial shift, met your divine helpers, begun a thorough magical hygiene practice, and been oriented to the concept of earthbound entities, you're more than primed to begin clearing your space! So let's continue.

When Do You Need to Clear a Space?

There are three primary reasons that will make it necessary and appropriate to clear a space. These are:

1: You're Moving In to a New (or New-to-You) Space

It's ideal to clear the decks before moving into a new space. This is because even if we have a general idea of what transpired in the space before we acquired it, we really don't know, especially from an energetic standpoint. And even if we did know, it's always a good idea to start out on the right foot by establishing the most positive vibration possible. Performing a thorough space-clearing ritual does just that.

2: Something Less Than Savory Has Transpired in the Space, and/or There Is a Significant Change in the Occupants

Naturally, a space-clearing ritual would be appropriate after an illness, a divorce, a disagreement, a roommate shift, a cocktail party gone awry, or any other occurrence that may cause you to want to refresh the space and reboot the energetic atmosphere.

3: You Suspect the Presence of Earthbound Entities

And then there will be times when you or someone you know suspects that an earthbound entity has entered or is just hanging around. (See previous chapter for indications.) Whether or not this is true, the unsettled feeling that caused the suspicion almost always can be successfully neutralized by a space-clearing ritual.

Periodic Clearings

In addition to the three main reasons stated here, it's a good idea to clear the space on a regular basis just because negative or stagnant energy, like dust, generally begins to accrue as a matter of course. Plus, space clearing is just plain fun! What's more, regular space clearings add an almost tangible, vivacious sparkle to the space and serve as a beautiful reminder that everything is magical and everyday life is sublime.

You might like to start off with a bang and perform the thorough, powerful, all-purpose space-clearing ritual below. This way you can get oriented to every dynamic and aspect of space clearing while continuing what you've started in the preceding chapters by infusing your life with a positive momentum and flow. Then you can perform less involved space-clearing rituals weekly or monthly (as you feel guided) to keep the energy fresh, such as a quick rattle or smudge as outlined in chapter 2 or any of the simpler rituals that appear later in this chapter, or, if you're feeling ambitious or like it's time for a biggie, you can always repeat this one or craft one of your own (more on this later in the chapter and in the appendix). As you continue to work with the ideas in this book, you will begin to have an intuitive sense of what type of clearing would be appropriate when.

A THOROUGH SPACE-CLEARING RITUAL

As clearing rituals go, this one is a great place to start. It's not only relatively simple and straightforward, it's also thorough, powerful, potent, and—provided you've sufficiently prepared yourself by reading and embarking on the exercises in the previous chapters—it's veritably foolproof. What's more, it's a trusty old standby ritual: regardless of the situation or your level of expertise, you can always come back to it and employ its amazing benefits.

I suggest that you start by performing it on your own home. (If you share a home with roommates or family members who wouldn't understand or approve, simply perform it on your bedroom or whatever space is exclusively yours. Or, if it's an emergency, do it when they're not home, avoiding their personal spaces as necessary or appropriate.) You can do it anytime of the day or night—whatever feels right. However, I do suggest performing it during the waning moon (when the moon is between full and new), as energies more naturally disperse and diminish during that time.

Feel free to use this book (no need to memorize).

Also, I suggest reading through the entire chapter before performing the ritual. This way, you'll have a better understanding of the significance of what you'll be doing before you actually do it.

YOU WILL NEED:

> Your rattle
>
> A bundle of dried white sage with dish
>
> A white or off-white soy candle
>
> A dinner plate
>
> 1 cup sea salt
>
> 12 copal incense sticks (frankincense will also work in a pinch) with dish
>
> A mister of spring water
>
> Essential oil of rosemary
>
> Essential oil of peppermint
>
> A pot of soil or a sealed container (to extinguish the sage and copal)
>
> Matches or a lighter

Before beginning, perform the clearing and shielding visualizations from chapter 4.

Also, open as many windows as you can. (Just opening them a crack is fine—and if weather is severe, leaving them closed will also be fine.)

One more thing: cover any open containers of food or drink, including pet food and water. This is because negative or stagnant energy that is released during the ritual can sometimes find its way into food and drink, sort of like dust or cobwebs falling from the ceiling.

Then, as preliminary prep work, add approximately twenty drops of essential oil of peppermint and forty drops of essential oil of rosemary to the mister of spring water. Close the bottle and shake. Hold it in both hands, and call on Archangel Michael (or a divine helper of your choice) to infuse it with clear, vibrant energy and very bright white light. Visualize/imagine/feel it pulsating and vibrating with this light.

Assemble all the ingredients somewhere convenient, such as your altar or kitchen table. Cover the dinner plate with sea salt. Hold the candle in both hands, and once again call on Archangel Michael (or a divine helper of your choice), this time requesting and visualizing that the candle be empowered with clear, vibrant energy and very bright white light. Just as you did with the mister, visualize/imagine/feel the candle pulsating and vibrating with this light. Set the candle in the center of the plate and light it.

Invoke the divine helper (or helpers) that you've chosen to work with for your space-clearing endeavors. Request their presence and assistance, and state your intention by saying the following invocation or something like it (but do hit all the main points):

> *Great Goddess, Archangel Michael, and all beings*
> *of harmony and light that are native to this space*
> *and surround this space,* I call on you!*
> *Please be here to guide and help with every aspect*
> *of this space-clearing ritual.*
>
> *It is my intention to clear out all stagnant and*
> *negative energetic conditions from this space.*
>
> *I now affirm, visualize, expect, and firmly establish*
> *utterly harmonious and healthy conditions*
> *within these walls, now and always.*

Thank you for being here.
Thank you for helping.
Thank you for making this ritual a success.
Thank you, thank you, thank you.
Blessed be. And so it is.

*Note: Feel free to change the names you invoke according to your personal helpers and spiritual orientation.

Now place the candle and plate near the center of the room you are in (it's okay if it's on the ground). Pick up your rattle and move around the room in a counterclockwise direction, shaking it loudly as you go. Set the rattle down and light the sage bundle. Again moving in a counterclockwise direction, as you hold the sage bundle (with a dish beneath it to catch any falling embers), let the smoke swirl around the space. Then place the sage in the dish and carry the sage/dish, rattle, and plate/candle into the next room or area. Set the candle near the center of the room again and repeat the rattle/sage process. Repeat this in each room and area of the home.

Return to the original location. Extinguish the sage and light the twelve sticks of copal. Repeat the process exactly as before, but with the copal this time. Again, do this in each room and area.

Return to the original location. Again, place the plate/candle in the center of the room. Extinguish the copal and pick up the mister. Shake it again, and then move around the room in a clockwise direction, misting the area as you go. Spray a bit of mist in the center of the room as well. Carry the plate/candle and mister into the next room or area and repeat. Repeat this process in each room and area.

Return to the original location. Hold your hands in prayer pose and visualize/imagine that the entire home is a sphere of very bright white light. You might ask for or invoke divine assistance with this. Request that this light transmute any and all negativity and stagnant energy into positivity, vibrancy, and blessings.

Now walk outside and place both your palms on the outside of your closed front door. Close your eyes and envision the door filled with vibrant, golden-white light. Say a prayer such as this:

> *Archangel Michael,* * *now that this space is clear*
> *and harmonized in every way, please allow only*
> *what is characterized by love—including vibrant energy,*
> *positive conditions, and beneficent beings—to cross this*
> *threshold and enter this space. Thank you.*

Repeat this threshold prayer with each doorway to the outside, including any garage doors.

Finally, say a prayer or invocation of thanks. Just a simple "Thank you so much, Great Goddess, Archangel Michael, and beings of harmony and light" will do, as long as you mean it.

Extinguish the candle. Dispose of the candle as responsibly as possible and flush the salt down the toilet. Wash the plate and then wash your hands.

Remember to uncover any companion animal food and water.

Note: Again, feel free to invoke the deity of your choice.

And Last but Not Least...

When you're finished, lie flat on your back for at least five full minutes (and possibly more if you feel guided or if your energy feels like it needs more grounding). You've just drummed up a bunch of extra energy that is now hanging around your energetic body. To return to a normal energetic condition that's more appropriate and healthy for everyday purposes, visualize/imagine/feel all the excess energy that was previously in and around your energetic body dripping down through the floor (and perhaps any stories beneath you) to be reabsorbed into the earth like water after a storm.

Then go in the kitchen and consume at least one of the following items or something similar:

- A beer
- 2 pieces of bread or toast
- A bagel
- A muffin
- A potato
- A bowl of rice
- A bowl of oatmeal
- 2–3 cookies
- A piece of cake

This will help ground your energy and bring you back into the physical realm.

Finally, take a shower or bath to further bring you into the physical realm while detoxifying and recalibrating your energetic field. Also, within one hour after the clearing, drink at least sixteen ounces of water to cleanse any toxins you may have absorbed into your physical body.

What's Really Going On During a Space-Clearing Ritual

Although you will almost definitely feel an unmistakable energetic shift after performing the above ritual, space-clearing rituals (and energy work in general) are mysterious and mystical by their very nature. Still, to a certain extent I'd like to demystify each step of the above ritual by describing what's going on from a metaphysical perspective. This way, it will be easier for you to begin to sense the subtle realm and subtle energetic conditions on your own. This will not only allow you to be a more effective space clearer, it will also enhance and reveal your psychic gifts and intuitive powers; eventually, it will allow you to feel comfortable experimenting in order to create and perform your own unique space-clearing rituals according to your needs and what you discover works best for you.

> **Invocation/Intention Declaration:** Because we are all one with everything, including the Divine, when you invoke a divine helper or helpers (or divine power in any form), you are tuning in to the part of you that goes beyond the illusion of separation. The true part of you—not your ego or the aspect of you that calls yourself Annie or Jacob or whatever—is the aspect of you that is one with everything and that has access to all information and all power. Once you align with

this power, when you clearly state your intention, it is like tuning a radio frequency to exactly the station you want so that you can be clear and direct with what you accomplish.

Candle: In many cultures and for many eons, lighting a candle has been a symbol that's helped humans to invoke and align with the Divine. Perhaps this is because a small, single point of flame has the power to start a raging inferno but instead is focused, serene, and contained. Similarly, when we invoke divine power, we are invoking something all-powerful but we are directing it toward a clearly defined point of focus. Additionally, candlelight helps open a doorway between the worlds by focusing our attention on the here and now while helping to relax our linear, conceptual mind.

Sea Salt: Sea salt is an invaluable tool for absorbing negativity. When left in the open on a plate while you are clearing a space, it serves to draw out energetic toxins from even the darkest and most remote corners of a room. It also neutralizes negative momentum to help clear the energetic slate of a room and allow more positive conditions to emerge.

Rattle: As we discussed in chapter 2, your rattle powerfully loosens and unsticks stagnant negative conditions. Additionally, beans are said to help earthbound spirits transition into the light and to persuade the more stubborn earthbound spirits to simply vacate the premises. The garlic in your rattle helps to clear away any sort of parasitic energy, i.e., psychic energy or energetic conditions that may deplete you in any way. When clearing, moving in a counterclockwise direction generally feels more powerful, because this is the direction that intrinsically loosens and unwinds (like in the phrase "righty tighty, lefty loosey").

White Sage: To review more material from chapter 2, white sage smoke improves the spiritual vibration of a space, purifies and disinfects it of energetic toxins, and helps transmute negativity into positivity. As such, it can also help open a doorway between this realm and the heavenly realm, aka the realm of light.

Copal: Copal opens up the doorway to the heavenly realm; it also infuses our space with sweetness and magic. Copal makes it very unlikely that stagnant energetic conditions or earthbound attachments will want to hang around.

Mist Potion: Once you've gotten through the bulk of the clearing, you might think of the mist potion as a spiritual disinfectant of sorts. Peppermint purifies the vibration and energizes the atmosphere, while rosemary blesses the space with a positivity that is sweet yet stubborn, pleasant yet firm.

Counterclockwise Direction: At this point in the ritual we move in a counterclockwise direction, because now, rather than clearing out the old, we are bringing in the new by consciously setting a fresh and intention-filled energetic tone.

Invocation or White Light Sphere: At this stage in the clearing you've just drummed up and set in motion some very positive energetic conditions. When you call upon and visualize the sphere of white light, you're setting the energetic conditions in place. In other words, you're protecting, preserving, and shielding them so that they'll stay fresh for as long as possible. You might also think of this sphere visualization like focusing and containing the positivity that you've just drummed up, sort of like spreading a beautiful blanket on the beach and then gently anchoring it down at the corners with rocks.

Threshold Prayer: Doorways between your home and the outside world are especially important when it comes to protection, as this is where everything (including energy, resources, and people) enters your home. After saying this prayer and doing this visualization, you'll further preserve the good vibes you've just established, and only positive energetic conditions, people that wish you well, and divine beings of love and light will be likely to enter your home. (But please still lock your door; for protective purposes, it's ideal to work with both spiritual and physical worlds simultaneously.)

Thank You: Always say thank you! Thanking the divine power and any beings that you have invoked is a way of honoring and showing gratitude to the all-loving, all-knowing divine aspect that resides within you and within All That Is. And when we show gratitude for something, it becomes more and more available to us in the present and future. This is why making sure to express a word of thanks after every ritual constantly increases the degree of metaphysical and spiritual power available to us.

Lying On Back: As stated above, when you perform a ritual and walk between the worlds, you call a lot of extra energy into your energetic body. Lying on your back and doing the recommended visualization helps you release all the extra energy and restore yourself to your normal state. If you were to skip this step, when you went back into your daily activities, it would be like trying to water tiny little seedlings with a gigantic firefighter's hose on full blast. Just as this would disturb the delicate nature of the seedlings, going about your daily tasks with this much energy would immediately undermine your inner equilibrium, and you could even throw your immune system out of whack and catch a little cold (like I did when I was a beginner who heedlessly skipped this step).

Eating or Drinking: Ingesting something grounding—such as a grain-based food, a beverage, or a root vegetable—can further help bring your energy back into balance. This time, it's less about removing the excess energy and more about reconnecting you to the physical realm. After all, you've just been walking between (meaning being aware of and consciously affecting) the physical and nonphysical realms, the latter of which most of us are not generally accustomed to. Reconnecting with the physical realm helps us to be steady, sure-footed, grounded, and mentally focused so that we can go about our daily activities safely and effectively.

Shower or Bath: Especially when performing thorough space-clearing rituals like the one above, and especially when you're new to energy work and space clearing, it's important to shower or bathe afterwards for a number of reasons. First of all, the feeling of water on your skin helps further bring you back into the physical, sensual realm. Second, the ions in the water help recalibrate your energy field by adding what is needed and removing what is not. Third, the water helps draw out any lingering negativity that you may have absorbed into your energetic field while clearing.

Water: Drinking water further helps to detoxify and recalibrate your physical and energetic bodies.

Additional Techniques

Once you've successfully performed your first space-clearing ritual (like the one above), you might like to experiment with some other techniques.

In this section, you'll find additional rituals that can powerfully clear the space. If you feel drawn to one of them and the time feels right, by all means try it! It's also fine to do one of the rituals outlined below instead of the above ritual for your first one if you intuitively feel drawn to do so.

Elemental Ritual

This ritual calls upon the power of the four directions and the four elements to powerfully cleanse a space.

YOU WILL NEED:

> Your rattle
>
> 1 large three-wick pillar soy candle
>   on a dinner plate
>
> A mister bottle filled with spring water
>
> 1 teaspoon sea salt
>
> 1 tablespoon fresh lemon juice or 10 drops
>   essential oil of lemon
>
> 2 cups non-iodized table salt in a ceramic bowl
>
> 12 frankincense incense sticks
>   stuck upright in a bowl of sand or salt
>
> A compass (or an innate orientation
>   to the four directions)
>
> Matches or a lighter

Preliminary Prep Work: Place the sea salt in the mister along with lemon juice or essential oil. Shake well.

More Preliminary Prep Work: Shake your rattle loudly as you move in a counterclockwise direction around the perimeter of each room and area.

Find enough floor space to designate an imaginary circle that's approximately six to seven feet in diameter. At the north edge of the circle place the ceramic dish of salt; at the east edge place the incense; at the south edge place the candle; and at the west edge place the mister.

When you're ready to begin, stand in the center of the circle, facing east. Take some deep breaths, relax, and center your mind. Light the twelve sticks of incense. Stand up straight, with your palms facing out and your arms slightly raised at about a 30-degree angle from your body. Say:

Powers of the east, spirits of air, I call on you!
Beings of sunrise, of the rushing wind,
of new beginnings and fresh ideas,
I honor you, I summon you, and I greet you!
Hail and welcome, guardians of the east!
I am grateful for your presence
and for your help in this clearing.

Now face south. Light the candle wicks. Stand up straight, with your palms out and your arms slightly raised. When your mind is appropriately focused, say:

Powers of the south, spirits of fire, I call on you!
Beings of noontime, of the summer sun,
of passion, courage, and joy,
I honor you, I summon you, and I greet you!
Hail and welcome, guardians of the south!
I am grateful for your presence
and for your help in this clearing.

Face west. Stand up straight, with your palms out and arms slightly raised. When you feel ready, say:

Powers of the west, spirits of water, I call on you!
Beings of sunset, of the ocean deep,
of emotions, fluidity, endings, and transitions,
I honor you, I summon you, and I greet you!
Hail and welcome, guardians of the west!
I am grateful for your presence
and for your help in this clearing.

Face north. Stand up straight, with your palms out and arms slightly raised. Say:

Powers of the north, spirits of earth, I call on you!
Beings of midnight, of the fertile soil, of stability,
nourishment, and sensuality,
I honor you, I summon you, and I greet you!
Hail and welcome, guardians of the north!
I am grateful for your presence
and for your help in this clearing.

Face east again and pick up the incense. Move around the room in a counterclockwise direction, letting the smoke waft around the area. As you do so, imagine a powerful wind of positivity sweeping through your home, and continually repeat these words: "With the power of air I clear this home." Repeat in each room and area, and then come back to the circle and return the incense to its original location.

Face south and pick up the candle. Move around the room in a counterclockwise direction. As you do so, imagine a fire sweeping through your home, burning away and transmuting all negativity. Also, continually repeat these words: "With the power of fire I clear this home." Repeat in each room and area, and then come back to the circle and return the candle to its original location.

Face west and pick up the mister. Move around the room in a counterclockwise direction, misting as you go. As you do so, imagine clear and frothy ocean waves washing through your home, powerfully purifying it of all negativity and leaving it sparkling. Continually repeat these words: "With the power of water I clear this home." Repeat in each room and area, and then come back to the circle and return the mister to its original location.

Face north and pick up the dish of salt. Move around the room in a counterclockwise direction, throwing small bits of salt against the walls and in the corners as you go. As you do so, envision the salt quickly neutralizing and absorbing any lingering negativity and the floor beneath your feet as cool, clean, receptive, nourishing soil. Continually repeat these words: "With the power of earth I clear this home." Repeat in each room and area, and then come back to the circle and return the dish to its original location.

Face east again. Say:

This home is now cleared.
By the powers of east, south, west, and north,
this home is now cleared.
By the powers of air, fire, water, and earth,
this home is now cleared.
The boundaries are set.
Henceforth, only good intentions, vibrant energy,
and beneficent beings may live within these walls.

Now it's time to thank the elements. To do so, while still facing east, say:

*Powers of the east, spirits of air, you were here
and you lent your strength, and for that I thank you.*

If it's still burning, extinguish the incense.

Face south. Say:

*Powers of the south, spirits of fire, you were here
and you lent your strength, and for that I thank you.*

Face west. Say:

*Powers of the west, spirits of water, you were here
and you lent your strength, and for that I thank you.*

Face north. Say:

*Powers of the north, spirits of earth, you were here
and you lent your strength, and for that I thank you.*

Face east again. Say:

Thank you, thank you, thank you. Blessed be. And so it is.

Important: Be sure to finish by lying on your back and then eating something grounding (see page 95 for suggestions).

Archangel Michael Candle Ritual

Even in the smallest California towns you can often find candle, incense, and herb shops called botanicas. At these botanicas you can invariably find very reasonably priced (less than $2) tall red jar candles adorned with a picture of Archangel Michael on one side and a prayer on the other. These are the candles I recommend for this ritual. If you don't have a botanica nearby, you can either find them online or substitute simple red candles in jars; if you feel moved to do so, you might also print out pictures of Archangel Michael and glue or tape them to the jars.

If you choose to perform this ritual, make sure to get your animal companions out of the house and make sure to be *very* mindful of fire safety.

YOU WILL NEED:

> 1 red Archangel Michael candle for each room
>   and area of the home
>
> 1 frankincense or copal incense stick for each
>   room and area of the home, along with some
>   way to hold each stick, such as an incense holder
>   or a small cup filled with dirt or sand
>
> 1 bundle dried desert sage

Place one candle and one stick of incense in a central location in each room and area. In the first room or area, light the candle and incense, and say:

> *Archangel Michael, I call on you!*
> *Please clear this space with your fiery light!*
> *Please escort any and all earthbound entities to the light!*
> *Please leave only love, positivity, and harmonious conditions*
> *in your wake! Thank you!*

Repeat in each room and area, then leave the house for an hour with your animal friends, if you have any. (You might want to hang out in the yard to keep an eye on things.) When you return, extinguish each candle, and each time you do so, say, "Archangel Michael, thank you!"

Discard the candles somewhere outside the house, clean up the incense situation, and wash your hands. Finish up by lighting the desert sage bundle and moving through each room and area of the house in a clockwise direction. This will reset and recalibrate the energy in a positive and comforting way.

Rainbow Woman Doorway of Light Ritual

As mentioned in chapter 3, Rainbow Woman can help transform your home, metaphysically speaking, into a shimmery doorway of sparkly light, drawing in and transmuting negative energy and entities and leaving highly harmonized energetic conditions in even the remotest corners of your home and etheric, opalescent sparkles everywhere you look.

YOU WILL NEED:

> 1 white or off-white soy votive candle and holder
> for each room and area
>
> 1 mister filled with rose water and 4 drops
> rainbow quartz elixir*
>
> 1 bundle dried white sage
>
> Matches or a lighter

*Note: Rainbow quartz elixir is available online or can be made at home. (For instructions on how to make gem elixirs/essences for household use, see my first book, *Magical Housekeeping: Simple Charms and Practical Tips for Creating a Harmonious Home*.) As an alternative to the elixir, just place one small rainbow quartz in the mister with the rose water.

In each room and area, in a fairly central location, place one candle with holder. In the first room or area, light the candle, and then say the following invocation:

Rainbow Woman, I call on you!
Please open up an irresistible doorway of light.
If any earthbound spirits are present within
these walls, please escort them through.
If any negative energy or stagnant conditions
are present within these walls, please transmute them
into positivity and vibrancy. Thank you!

Repeat in each room and area.

Then return to the first room. In a counterclockwise direction, move around the space while misting with the rainbow quartz/rose water mixture. Repeat in each room and area. Being very careful of fire safety, allow the candles to continue to burn for one to two hours. (Staying in the house is fine.)

Then, while the candles are still burning, light the white sage bundle and waft the smoke around each room and area while moving in a counterclockwise direction. Extinguish the sage.

To conclude, stand in a central location and say:

Rainbow Woman, I thank you for your presence and
your powerful assistance! Now that this home is clear
and vibrant, please close the doorway of light, but may
the positivity and blessings that it brought remain
in its wake. The boundary is set; this home is clear!
Rainbow Woman, I thank you!

Extinguish the candles and discard them.

Basic Space-Clearing Ritual

As I mentioned earlier in the chapter, sometimes you can sufficiently clear the space with just a quick rattle or smudge. Other times you'll be in the mood for an intense ritual like any of the ones above. But perhaps most times you'll want to perform a thorough, effective clearing without spending too much time or energy on fancy prep work, incantations, or visualizations. And for those times, this informal ritual (or a variation on the theme) would be the one for you. I like to do it after I clean the house to clear the energetic decks and add an etheric glimmer and breath of magic to my already sparkling home.

YOU WILL NEED:

> Your rattle
>
> A bundle of dried white sage (with dish)
>
> Matches or a lighter
>
> A sealed container or dish of sand (to extinguish the sage)
>
> A mister of rose water into which you've added 4 drops Rescue Remedy and 8 drops essential oil of lavender

Remember to cover any animal food and water beforehand, and then uncover it when you're finished.

Begin by shaking your rattle loudly while moving in a counterclockwise direction through each room and area. Light the bundle of sage. While holding a dish under it to catch any burning embers, again move in a counterclockwise direction through each room and area, letting the smoke swirl throughout the space. Pay special attention to corners and anywhere where you imagine the energy might get stuck. Extinguish the sage.

Shake the mister and, as you move in a clockwise direction, mist the perimeter of each room and area. Finally, stand in a central location or just inside the front door (facing in toward the house). Hold your hands in prayer pose and envision a very bright sphere of golden-white light completely encompassing your space. To seal the positive energy in place, see it moving in a slight clockwise direction. During this time, you might set your intentions for the space and send them through this light. For example, you might think the word *love* or *prosperity* and mentally send the echo and vibration of this word through the visualized light.

VARIATIONS ON THE BASIC SPACE-CLEARING RITUAL
Instead of the white sage, you might use:

- A bundle of dried desert sage (to create a feeling of comfort, move the energy around in a playful way, and open doors of possibility in your life)

- A bundle of dried mugwort (to create opportunity, call in blessings, and protect the space from negativity)

- 3–5 sticks of frankincense incense (to establish a very high spiritual vibration and align the space with the energy of heaven)

Instead of the rose water mixture, you might use:

- Spring water with 4 drops white chestnut essence and 8 drops clary sage essence (to create

clarity and minimize excessive thoughts and
worries)

- Rose water with 4 drops crab apple essence
and 8 drops ylang ylang essence (for sensuality,
romance, self-love, and self-acceptance)

- Spring water with 2 drops aspen essence,
2 drops star of Bethlehem essence, and 8
drops essential oil of neroli (for peace, family
harmony, healing, grounding, and freedom from
fear)

- Spring water with 4 drops honeysuckle essence
and 12 drops essential oil of eucalyptus (to get
the energy moving in a healthy way, promote
physical healing, and promote motivation and
positive change)

Instead of the white light visualization, you might visualize:

- Green light with gold sparkles (for abundance)

- Pink light with etheric rose petals (for romance)

- Blue light (for protection and peace)

- For more ideas, see chapter 9

When You're Ready to Craft Your Own Rituals

In appendix A, you'll find a listing of ingredients that will come in handy when you're ready to begin crafting your own rituals. If you feel comfortable doing so, you might also try substituting ingredients from appendix A in any of the above rituals according to your intentions and needs (and according to convenience—i.e., what's available at nearby stores or what's already lying around your house). Just be sure to read the description of the ingredient carefully and to have a good understanding of the metaphysical workings of a space-clearing ritual (see page 96 for more on this topic) so that you can substitute safely, appropriately, and effectively.

seven

CLEARING OBJECTS
AND PEOPLE

imply loving your items and the people in your life infuses them—and you, by virtue of your relationship with them—with positive energy. Furthermore, if you're regularly engaging in the magical hygiene practices from chapter 4 and the space-clearing exercises from the previous chapter, you've already got your good mojo flowing in a powerful way. Still, there can be certain instances when clearing an object, a loved one, or yourself might be something that you want to do—for example, if something about a secondhand item isn't feeling totally copacetic to you, or if you or your partner just came home from a challenging situation (such as a hospital visit) and you feel a little "off," or not entirely like yourself.

In these situations, performing a clearing on an object or person can sometimes be helpful, because thoughts, emotions, occurrences, and situations have energetic charges, and this energy can become lodged within and begin to characterize the energy fields of people and items. Most of us have an innate understanding of this, which is why we would not want to use a murder weapon to cut our vegetables no matter how many times it had been cleaned and disinfected, and why we'd be likely to consider a home run baseball lucky or at least valuable in some way. Similarly, the popularity of family heirlooms and museum relics proves that we, as a culture, believe that we can glean a "certain something" of the personality or spirit of the previous owner by living with or gazing at their items.

Additionally, in certain rare cases, it's said that an earthbound spirit can attach itself to an item or person that may be an energetic match for them in some way (such as a wedding ring the spirit used to own or a person who suffers from an addiction that the spirit used to have). If you suspect something like this may be going on, needless to say, it's also a good idea to clear the object or person as necessary.

Please note: I don't want this—or anything in this book, for that matter—to be one more thing for you to obsess over. Rather, I want to help make you more aware of the subtle interplay of energies in your life and your home, and to have the tools you need to affect them for the better, if and when you feel that it will be helpful. After all, the most important things are your intention and your mental and emotional well-being, so attend to those things first (or allow your clearing work to be a means to those ends) by remembering that you are safe, you are spiritually guided and protected, and all is well. Then you can interact with both physical and spiritual realms from a place of grounded strength rather than worry or panic. And, in turn, this will positively affect every aspect of your life.

WHEN DO YOU NEED TO CLEAR AN OBJECT?

There are a number of situations that call for the clearing of an object.

New-to-You Items

Almost anytime you purchase or obtain a secondhand item of any sort—an article of clothing or jewelry, a piece of furniture, a decoration, or anything else—since you usually can't know for sure where it's been or what sort of energy it's been subjected to, you'll probably want to perform some sort of clearing on it.

There are a couple of exceptions to this rule. Namely, you don't need to clear a secondhand, new-to-you item if:

- You received it as a hand-me-down from a trusted loved one (living or deceased) whose energy you treasure and would prefer to preserve within the item.

- Your intuition tells you that its energy is 100 percent positive, and you would prefer to preserve this existing positive energy within the item.

Secondhand Items You Already Own

If you already own a secondhand item that you suspect may be holding negative energy, it's a good idea to clear it. Many (if not most or all) of your belongings should be sufficiently cleared by default when you perform a clearing for your home, but if you have a feeling that an item needs a little something extra, by all means follow your intuition.

Items Created or Sold in an Exploitative or Otherwise Negative Way

I have often seen beautifully carved wooden Buddha statues for sale that have a very sad feeling about them. I believe this is because the people who carved them were not happy or treated well. Now that you're learning to tune in to and trust your intuition, you'll probably find yourself having similar intuitions. But in the meantime, if your intuition tells you that you're living with any items that are like the sad wooden Buddhas—perhaps a basket that was woven by a mother worried about where her children's next meal was going to come from, or an indoor fountain that you purchased at a store where the owner was always in a terrible mood—you'll want to either donate the item to charity (which will help transmute it into a more positively vibrating item by lending its energy to a good cause, especially when you release it with this intention) or perform a clearing on it.

Items that Hold Negative Energy or Old Associations from Your Own Past

And then there is the case of the bed you once shared with your ex-husband or the dining table where you were sitting during that memorably tragic meal. If now's not the time for a brand-new bed or dining table, you will definitely want to perform a clearing on the item in question. While this might not entirely remove the negative energetic charge it holds in your memory, it can still do a lot by helping you clear the energetic residue, reboot the energetic pattern, and reframe your relationship with the item.

Even if the memory isn't completely devastating, you might still want to clear the item. Take, for example, the bed you once shared with your ex. The relationship wasn't terrible, and you are even still on good terms—for one reason or another, it just didn't work out. Still, the bed powerfully holds the energy of the old relationship. This means that it might be energetically blocking a new relationship. In other words, it might be keeping new romances away because, on a subtle level, sensitive potential love interests get the feeling that you're already attached. Even if this isn't the case, any new relationships you attract will almost definitely, at least in some ways, repeat the tired, unwanted patterns of the old one.

How to Clear Objects

Because of the diverse nature of the objects that you might be clearing, there are a number of different techniques you may want to employ. But, to begin, here's a general ritual you can tailor to fit most objects.

A General Object-Clearing Ritual

Step One: If possible and appropriate, physically wash the object thoroughly. (As always, if you employ cleaning solutions, choose products that are as natural and environmentally friendly as possible.)

Step Two: Place the object on a clean white cloth. (This is mainly for small and medium-sized items. If the item is very large, like a car or a sofa, don't worry about this step.)

Step Three: Set the object (and cloth) in bright sunlight for anywhere from five minutes to an hour. (While this is ideal, if it is challenging or impossible for any reason, it's fine to skip this step.)

Step Four: Light a bundle of dried white sage. Smudge the object by allowing the smoke to generously waft around it and saturate it as much as possible. If it's an object that opens up in some way, such as a wardrobe or a vehicle, open it up and smudge the inside as well.

Step Five: Say a prayer or blessing such as this:

> *Archangel Michael, please cleanse this _____*
> *(ring, hat, table, etc.) of any and all stagnant energy*
> *and energetic attachments, and please infuse it with*
> *very clear, vibrant bright light and positivity.*

Great Spirit, please bless this _____ and align it with the
truth of everything, which is infinite love and pure divinity.

I ask for and affirm that at this moment and in the future,
> *this _____ is made of divine love, surrounded*
> *with divine love, and employed solely for*
> *divine and loving purposes.*

Thank you, thank you, thank you. Blessed be. And so it is.

Step Six: Visualize the item completely filled with, made up of, and surrounded by a sphere of blindingly bright golden-white light.

ITEM-SPECIFIC CONCERNS AND TECHNIQUES

While the above ritual will work well for almost any item, here you'll find some additional item-specific concerns and techniques.

> **Clothing:** To clear the energy of multiple items of clothing, add one-half cup of white vinegar to the wash cycle. If you're also drying the item(s), put five drops of sage, cedar, or sandalwood essential oil on a small scrap of cotton fabric and place it in the dryer with the rest of the clothes. Then perform the fifth and sixth steps from the general object-clearing ritual.

> **Beds and Upholstered Furniture:** Cloth furniture is especially energetically absorbent, so if you know something negative happened on a bed or an upholstered piece, or if something occurred on it that you don't want to repeat (such as discord from a former relationship), I recommend getting rid of the piece altogether and replacing it with a new one. Still, there will be times when it's necessary to clear cloth furniture, such as when you find or inherit a secondhand piece that you love or after a break-up when it may not be an ideal time to purchase new furniture. In some cases, reupholstering is a great option. Otherwise, wash the item as best as you can, then perform the fourth, fifth, and sixth steps from the general object-clearing ritual. Finish by strewing fresh rose petals across the top of the item. Leave overnight. In the morning, remove the petals

(place them near the base of a tree or in a compost bin or heap), repeat the fifth and sixth steps, and lightly spray the item with a mister of rose water into which you have added forty drops of sage, cedar, and/or juniper essential oil (in any combination).

Paintings and Other Artwork: Artwork is like an extension of the artist and powerfully holds the energy of whatever emotions the artist was experiencing during the time of creation. This means that it's important to tune in to the feeling the piece gives you and to make sure that it's a feeling you're comfortable bringing into your home. Still, it's a good idea to incorporate an aspect of transformation into your cleansing ritual, so that any negative energy that may be trapped within the piece can be transmuted into positivity and love. To do this, perform the general object-clearing ritual exactly, only instead of the prayer/blessing in step five, say something like this:

Archangel Michael, please clear this _____
(statue, painting, etc.) of all negativity and negative
attachments, and please infuse it with the bright light
of positivity and love.

Saint Germain, please transmute and transform
all energies in this _____ into positivity and love.
May this benefit the artist and all beings
in all directions of time.

*Great Spirit, please bless this _____ and align it
with the energy of pure love and pure divinity.*

*May all who gaze upon it be uplifted and blessed.
Thank you, thank you, thank you. Blessed be. And so it is.*

Mirrors: Mirrors are among the most mystical of objects,
not the least because they can absorb the energy of the
situations and environments they reflect. And so, unless
you know where it's been or you have a very reliable,
positive intuitive feeling about it, it's often a good idea to
clear secondhand mirrors thoroughly or, in some cases,
to steer clear of them altogether. And if you find that
a mirror-clearing ritual is in order, simply perform the
general ritual above, making the following alterations:

During the first step, wash the surface of the mirror
first with ocean water or spring water mixed with sea salt
(approximately one tablespoon salt per cup water), and
then again with white vinegar. (The white vinegar will
act as an effective glass cleaner.)

Do not skip step three! But do be careful not to start a
fire or to blind yourself or anyone else.

In step five, instead of the first line as listed above,
add the italicized words below by saying: "Archangel
Michael, please clear this _____ (ring, hat, table, etc.)
of all negativity and negative attachments; *please cancel,
clear, delete, and erase any old situations, environments, and/
or memories that it may hold*; and please infuse it with the
bright light of positivity and love." Then say the rest of

the invocation and finish the remainder of the ritual as recommended above.

Crystals, Rocks, and Gemstones: While it will never hurt to perform the full general ritual (above), usually the energy of a crystal, rock, or gemstone can be effectively cleansed and reset simply by performing steps one through four.

Dishes and Cutlery: If you obtain a set or a large number of secondhand dishes and/or cutlery and you want to clear them all at once, wash them with soapy water (using all-natural dish soap) into which you have added a good portion (one cup or so) of sea salt. Rinse thoroughly. Then perform steps two, three, and six as above. The white cloth may be a sheet or a beach towel, and the visualization can include the entire group of items at once.

Animal-Derived Items (such as pearls, coral, seashells, silk, wool, leather, feathers, and fur): Generally speaking, animal-derived items are not created with respect or kindness and therefore most often contain the energy of pain, suffering, and cruelty. (This is, of course, not true for things like naturally shed feathers or found seashells.) As we've seen, from an energetic perspective like attracts like, so you might want to think twice about bringing animal-derived items into your home. Still, if an item is secondhand and/or you already own it, the damage has already been done, so if you love the item and still want to live with it, it's ideal to perform a clearing that includes the aspects of (a) gratitude toward the animal and (b) transformation and transmutation of suffering.

To do this, perform the above general ritual exactly, only instead of the prayer/blessing in step five, say something like this:

Archangel Michael, please clear this _____ (chair, blanket, string of pearls, etc.) of all negativity and negative attachments, and please infuse it with the bright light of positivity and love.

From the bottom of my heart, I express compassion and gratitude toward the (cow, sheep, oysters, etc.) from which it came.

Saint Germain, I request and affirm that you now heal any negative feelings or energies in this (item) that the (source) may have left behind.

Please transmute and transform them completely into positivity and love. May this benefit and bless the [type of animal], along with all beings in all directions of time.

Great Spirit, please bless this [item] and align it with the energy of pure love and pure divinity.

Thank you, thank you, thank you. Blessed be. And so it is.

CLEARING YOURSELF
(OR SOMEONE ELSE)

Sometimes you'll find it helpful to go above and beyond the self-clearing aspects of a good magical hygiene practice (see chapter 4) by performing a self-clearing ritual. It can often be just the thing to reboot your emotions and energy field, infuse you with clarity, and place you fully and squarely into the type of mindset that will lend harmony and success to every aspect of your life. Periodic self-clearing rituals, or self-clearing rituals performed as needed (see below), can be especially helpful for spiritually sensitive people like us, especially those of us who go out of our way to sense and work with the subtle realm.

Times when you might like to perform a self-clearing ritual include:

- When you intuitively feel a bit "off"
- When you feel like you might have picked up someone else's emotional "stuff" (For example, I often feel this way after I perform feng shui consultations or intuitive counseling sessions— which brings us to the next point …)
- After performing energy work or healing for others
- After a challenging period or situation
- After being in a crowd or around a large number of people
- When you feel drained by a person or a situation

- When you feel inexplicably drained
- Every now and then (for fun and good measure)

Note: Once you get the hang of clearing yourself, you can move on to helping to clear other people if they request it. But for now, I'm going to teach you how to clear yourself. Then, when you're ready to clear other people, you can simply apply the same techniques.

A Thorough Self-Clearing

When a thorough self-clearing is in order, here's a great one you can do.

Obtain a large water bottle with a lid. Fill it with drinking water. Hold the bottle in both hands and visualize very bright white light filling the bottle. Say a quick invocation silently or aloud, such as this: "Great Goddess, please fill this water with vibrations of purification and love." Throughout the remainder of the clearing, keep this water handy and take a swig whenever you think of it and whenever you feel the least bit thirsty.

Soak in warm bathwater—into which you have dissolved at least one cup of sea salt—for at least forty minutes, adding hot water to the tub as necessary to keep the water comfortable. Before you get in the bathtub, visualize very bright white light filling the water. Say a similar invocation to the one recommended above. (If you don't have a bathtub handy, substitute with the power shower on page 129.)

After you've dried off, light a bundle of dried white sage and smudge yourself by letting the smoke surround and saturate your body and energy field.

Put on some clean, comfortable clothing, and sit comfortably, with your spine straight.

Say a prayer or invocation such as this:

Archangel Michael, please purify and detoxify my energy
field in every way, and, with your bright, fiery light,
please infuse me with vibrant positivity and love.

Saint Germain, please purify and detoxify my energy field
in every way, and, with your bright violet light, please
transmute all energies into joy and all forms of blessings.

Great Goddess (or Great Spirit, etc.),
please shield and protect me in all ways.
Thank you, thank you, thank you.
Blessed be. And so it is.

Visualize a sphere of very bright white light completely filling and surrounding your body and energy field, energetically cleansing and disinfecting you from the inside out.

Visualize a sphere of very bright violet/indigo flame filling and surrounding your body and energy field, transmuting and transforming negativity, recalibrating your energy field, and sealing and protecting you in all ways.

SIMPLE AND QUICK SELF-CLEARINGS

When you don't have time for a hot bath or you only need a quick energetic freshening-up (rather than the whole enchilada), a simple, quick self-clearing might be just the thing. And, because the power of energy work really has to do with two things—your laser-focused intention and your ability to tap into and align with the divine energy source—once you get the hang of clearing yourself, you might find that you can easily pack the full effects of a thorough clearing into any of the simple clearings below.

Power Shower: Moving water generates plenty of negative ions, which is literally the scientific term for good vibes. Before you take your normal shower, say a blessing invocation to the water such as this: "Great Spirit, please infuse this water with vibrations of positivity and purification." Then visualize the water as very bright white light that washes around and through your entire body and energy field, clearing away stagnant energy and negativity and transmuting it into vibrant energy and positivity.

Incidentally, there's no reason why you can't do this every time you take a shower.

Protective Potion: Obtain a mister filled with rose water. Add four drops of crab apple essence and four drops of white quartz essence. Shake the mister and visualize very bright white light coming down from above, entering the crown of your head and going down to your heart, through your arms, out through the palms of your hands and into the rose water. See it glowing and overflowing with a sphere of light that resembles a miniature sun. Say: "Great Spirit, please infuse this rose water with vibrations of purification, positivity, and love." Mist your entire body and energy field with it. (You only have to do the invocation/visualization once, and then the remaining mist is good to go.)

This is a great one to take on the road—you might even get a tiny mister and carry it along in your purse to use as spiritual first aid.

Sage: Simply smudging your energy field with a bundle of dried white sage can successfully detoxify your energetic field.

Archangel Michael Invocation: At any time, whether you're alone or in public, you can call on Archangel Michael to clear your energy field. Simply say or think these words: "Archangel Michael, please clear me of all negativity and all negative attachments. Thank you." Then you might visualize him removing negativity and attachments from you with a glowing vacuum tube (as author Doreen Virtue suggests). Then see him surrounding you with a sphere of very bright white light that further dissolves and transmutes all negativity into positivity. Finally, envision the sphere of white light turning into a sphere of very bright indigo/violet light that protects and shields you from all negativity.

Saint Germain Invocation: At any time, whether you're alone or in public, you can call on Saint Germain to surround you and fill you with his violet flame. Then say or think these words: "Saint Germain, please powerfully cleanse and incinerate any and all challenging energies in my body, mind, spirit, and energetic field, and transmute them into love and well-being. Thank you."

You can then visualize his violet flame. See it filling and surrounding you, quickly replacing all negative, stagnant energy with positive, vibrant energy.

White Light Vacuum and Sphere: At any time, whether you're alone or in public, you can call on the white light itself to vacuum you of all negativity. Simply take a deep breath, relax, and consciously choose to connect with the part of you that is eternal and that is one with the Divine. From this place, visualize a glowing vacuum tube of white light removing all stagnant and negative energy from your entire body and aura. Then visualize a glowing sphere of white light completely filling and surrounding your body, transmuting any remaining negativity into positivity and love.

eight

ENERGETIC PROTECTION TECHNIQUES

ou are an immensely spiritually powerful being. In fact, you are a sorceress (or a sorcerer), and you have within you everything you need to create the conditions you desire.

Before you go any further, stop. Relax. Take a deep breath or two. Now reread the above two sentences once more, slowly. Feel them in your body, and inhale them deeply into your awareness and belief system. Own them. You might even look in the mirror and confidently affirm: "I am an immensely spiritually powerful being. In fact, I am a sorceress (or sorcerer), and I have within me everything I need to create the conditions I desire."

Before I discovered that this was true for me, and before I had even been properly oriented to the unseen realm, I had a sense that there was a lot going on that I couldn't see and had no control over. As a result, my life experience was usually filled with panic, anxiety, and depression. By the same token, I was unwittingly allowing my spiritual power to fly all over the place, which often created exactly the conditions I most feared.

However, once I knew I was a sorceress, and once I discovered the tool with which to harness my spiritual energy—which is, of course, my laser-focused intent—I realized that I had the power to channel my inner dialogue and expectations so that I could establish the conditions I wanted to experience, rather than the ones I didn't. And, of course, so do you!

For example, when we feel fear around a particular situation or issue, instead of going down the slippery slope into panic—and, in the process, letting our power get away from us—we might confidently call on angelic protection and visualize a sphere of white light around us.

This isn't to say that we need to judge or berate ourselves for our natural reactions to stressful conditions, but it is to say that we'll have more success and experience more positivity in general when we deal with worry and fear from a place of empowerment rather than one of disempowerment. To put this another way, once we become conscious of our present inner dialogue—i.e., by acknowledging our thoughts and feeling our feelings—we can then begin to channel the flow and shift the tide so that we are using our considerable mental powers to nurture and preserve the most positive conditions possible in our personal space and all areas of our life … and that's what energetic-protection techniques are all about.

Protecting a Space

If you've performed the first thorough space-clearing ritual from chapter 6, you've already sealed your space with powerfully protective energies. But there may be times when you'll want to refresh or fortify your home's protective force field, perhaps with a thorough protection ritual (like the one below) or a quick space-protection cure (like one of the rituals that appear a little later on in the chapter). And, if you're like me, you'll want to incorporate a simple yet powerful protection visualization (which you'll also discover below) into your regular meditation practice.

A Thorough Protection
The Fiery Wall Ritual

As I mentioned above, the thorough space-clearing ritual in chapter 6 includes powerful protection. Still, if you want to infuse your protective efforts with a little extra oomph, you might follow up with this ritual. Or, if your space already feels pretty clear and you just want to shroud your home (and life) in an aura of protection, you might like to perform the following ritual after quickly refreshing the energy with a simple smudge, visualization, rattle, or prayer.

This is also a great one to do anytime you feel unsafe in your home for any reason.

YOU WILL NEED:

> ¼ cup olive oil
>
> ¼ teaspoon cayenne powder
>
> 1 clove garlic, pressed or minced
>
> A small bowl
>
> A tealight candle for every door and
> substantial window to the outside
>
> A clear jar or glass for each candle
>
> A long lighter or long matches

Mix the olive oil, cayenne pepper, and garlic in the small bowl. Hold your hands over the bowl and direct fiery white light through your palms into the oil. Say:

> *I now charge this oil with*
> *the bright, fiery light of protection.*

Assemble the candles and repeat the charging process by holding your hands over them and directing white light into them. Say:

> *I now charge these candles with*
> *the bright, fiery light of protection.*

One by one, remove each tealight candle from its aluminum container (if it has one) and use your finger to anoint the entire surface of the candle, excluding the wick, with the oil. If you like, you can place the candle back in its aluminum sleeve. Then place the candle in the jar.

When all the candles are prepared, set them on the floor just inside each door and on each windowsill (or near each window if there is no windowsill). In a generally clockwise direction, walk around and light them. Say:

As these candles burn, a fiery wall of protection is cast.
Inside this wall only love remains.
Through this wall only love may enter.
The wall is strong, the wall lives long, the wall burns bright.
A fiery wall of protection is cast.

Allow the candles to continue to burn until they are all burned out. Then recycle or discard the aluminum in which they came, and wash the jars or glasses thoroughly. You can discard the remainder of the oil or put it in a bottle or jar and save it for use in the fiery oil cure on page 139.

Daily Home Clearing and Shielding Visualization

I suggest incorporating this and the following invocations and visualizations (or ones like them) into your daily meditation practice. Please feel free to substitute any deities or representations of the Divine that feel powerful to you for the purpose. You can also feel free to call on or envision divine energy that is non-deity oriented, such as Light.

Close your eyes and relax.

Ask Archangel Michael to clear your home of all negativity with a glowing vacuum tube of light, and visualize him doing so.

Ask Saint Germain to transmute any and all negativity into positivity and blessings with his violet flame. Visualize this violet flame encompassing your entire home in a giant sphere.

Request that a circle of angels surround your home in a giant circle facing inward, preserving the positive energy in your home and directing inward the energies of love, light, clarity, serenity, and peace.

Request that another, larger circle of angels surround the first circle of angels. Ask them to face outward and send all negativity back to its source. For extra credit, you might also ask them to additionally send along exactly the energies that are most needed to transform and transmute this negativity into positivity and blessings.

Feel confident that these angels are trustworthy and that you are completely safe and divinely protected in all ways.

Additional Home-Shielding Visualization (for Overachievers)

Once the above visualization becomes thoroughly easy and quick, or if you're an overachiever by nature, you might want to follow up the above visualization with this one. It further protects your home from above and below and preserves positivity with the energies of earth and cosmos.

Visualize your entire home being encompassed by the trunk of a gigantic tree. The perimeter of this trunk extends outward from and completely contains the perimeter of your home.

See the roots of this tree descending deep into the earth. Watch them go down and down until they enter the molten core, which is filled with blindingly bright, golden-white earth light. See the trunk rooted firmly in the earth, drawn downward by this magnetic light. See the roots drink this light upward until it reaches your home, filling the entire interior.

See the top of the tree grow up and up, the branches reaching toward the sky until they extend outside of our atmosphere and into the sparkly clear cosmic light.

Let the branches drink in this light, and see it move downward until it reaches your home. See the golden earth light and sparkly clear cosmic light merging and mixing within your home.

QUICK HOME-PROTECTION CURES

Fiery Oil Cure: Once you've made the oil suggested in the previous fiery wall ritual, you can use it to anoint the outside of all the doors to the outside. To do this, simply use your finger to put a dab of oil in the center of the door, the middle top of the door frame, and the middle of each side of the door frame. Then say a quick prayer/ invocation to Archangel Michael, such as this:

Archangel Michael, thank you for powerfully protecting this home with a fiery wall of protection. Through this door, now and always, only love may enter.

Garlic Cure: Peel a clove of fresh garlic and cut it in half. Use the juicy side to invisibly write the rune Ansuz on the outside of each door to the outside. Ansuz looks like this: ᚨ

As you draw each rune, or just after you draw each rune, say:

Great Spirit, I call on you to protect this home.
Thank you.

If it feels right to you, instead of the rune, feel free to employ a protective symbol or sigil from your own spiritual tradition, such as a cross, a pentacle, or a Star of David. Afterwards, bury the garlic or place it in a compost heap.

Beans Cure: Hold a small handful of dried white beans in the sunshine, letting them bask in the light. Say:

I call on the light of the sun to fill these beans
with fierce positivity and the fiery light of protection.
I call on the spirit of the beans to send all
earthbound entities away or to sing them
gently to the light. Thank you.

Place a few beans near the outside of each door to the outside. (This one is especially good for when you have recently cleared a spirit or spirits from the house and you want to make sure that they don't reenter and that no new ones show up.)

Etheric Pentagram Cure: This is a great one to do if you have no tools at your disposal other than your intention and focus. But be aware—it does take lots of both!

Stand outside the front door. Say:

I call on the powers of the elements
—earth, air, fire, and water—
to protect this home from all harm.

Extend the index finger of your right hand and direct a laser beam of golden-white light through it. Starting in the top center of the star, draw a five-pointed star on the door by moving down to the lower right corner of the star first, up to the middle left corner, across to the middle right corner, down to the lower left corner, and back up to the top center. Then draw a circle around the star in a clockwise direction. Say:

Through this portal and into this home
only good may enter. It is blessed, it is sealed, it is safe.
Thank you, thank you, thank you.
Blessed be. And so it is.

Repeat at every door to the outside. It won't hurt to do the windows, too, if you feel inspired.

Aud Guray Cure: This is a highly protective chant from the Kundalini yoga tradition and another great protection to do if you don't have any physical tools to work with. The chant incorporates a number of ways of saying "I bow to divine wisdom," and it helps override your ego in order to channel your own divinity and your own divine ability and right to protect and be protected.

In addition to its meaning, its power comes from the vibration generated by its sound. Stand outside the front door and place both your palms flat against the front door. Mentally channel the energy of the chant into the door as you say:

AD GurAY NamAY

JugAD GurAY NamAY

SAT GurAY NamAY

Siri Guru DAY VAY NamAY*

* Note: This is spelled phonetically.

Repeat three times. Repeat at each door and window to the outside. (You can stand inside if the windows are upstairs or you can't get to them outside.)

ADVANCED PERSONAL PROTECTION TECHNIQUES

In chapter 4, you learned how to energetically shield yourself on a daily basis. Now it's time to learn a bit more about the dynamics of personal protection. These techniques apply to energetically protecting both your home and yourself.

Accept and Love What Is

When I first began working as a professional feng shui consultant, I would find that consultations would completely drain me of energy. I assumed that this was because I was opening myself up to other people's negativity and the negativity in their homes. In time, however, I learned that it was not so much my clients and their homes that were draining me but the degree to which I resisted the emotional and energetic patterns at work.

In other words, when I said an inner no to the conditions at work in their homes and lives, I was pushing against the truth of the moment, which, needless to say, took a lot of energy. When I instead said an inner yes, I left in a much more positive and energized state.

This might seem confusing. After all, wasn't I there to tell them what was wrong so that they could fix it? Well, not quite. While I was there to give them advice and help manifest the positive change they were craving, there was nothing actually wrong. I found that the perspective that was the most helpful to me, my clients, and the overall success of the consultation was that everything was perfectly unfolding in precisely perfect ways and in precisely perfect timing. The past state of their home, just like the present experience of the consultation, the future state of their lives and affairs, and my role in helping them get from here to there, were all essential pieces of the same whole, perfectly unfolding pie.

How does this apply to you? As much of a paradox as it may seem at first, it's important to cultivate a perspective that where you are now—emotionally, spiritually, and physically—is by necessity exactly perfect. This is the mindset that will allow you to traverse any situation with your positivity intact, and it will also allow you to fully open yourself up to each lesson, each kernel of wisdom, and every positive shift that your spirit craves.

To illustrate this, imagine that you're taking a road trip from Los Angeles to San Francisco. Now imagine that the whole time you're in the car you just keep thinking, "No, this place isn't right, because it isn't San Francisco." Then, once you get to San

Francisco, you keep thinking, "No, this isn't how San Francisco is supposed to look, because it isn't like any of the places I always see in movies." Can you see how this would close you off from the truth and the spontaneity of the moment, not to mention defeat the whole purpose of a road trip? If, on the other hand, you fully allowed each moment and said an inner yes to every experience, then every seemingly nondescript truck stop along the way could become in some way delightful, and only then could every park and street corner in the city become an opportunity to soak in the sights.

Additionally, when you encounter other people, places, and situations, remember that they are a reflection of you and you are a reflection of them. As such, love and accept yourself by loving and accepting every single thing that you encounter to the very best of your ability. You will soon discover that this is actually the attitude that will facilitate positive change in the easiest and most pleasant of ways.

Release Judgment

The idea of releasing judgment is really just another way of saying "accept and love what is." But it's important to approach this concept from a number of angles, because now that you're opening yourself up to the subtle realm and your intuition more and more, there is likely going to be a tendency to pass judgment harshly, such as "That man has bad energy, so he's a bad person" or "The vibes were so terrible in that place—those people ought to be ashamed of themselves."

Remember: we are all one. Everyone and everything you encounter is a reflection and actual representation of you. So, just as passing judgment on places, situations, and other people is the same as passing judgment on yourself, loving and accepting places, situations, and other people is the same as loving and accepting yourself.

Again, it might seem like a paradox, as in, "If everything's so perfect, why do we need to clear at all?" Think of it like this: just as you might lovingly notice that it's time for your child to take a shower after camping in the dusty desert or that it's time for you to brush your teeth after eating a spicy meal, you might lovingly notice that a place or a person could use a bit of energetic clearing.

Similarly, if you think of the earth plane as a sort of school or training ground (as I do), you might think of it like this: just as there would be no problem with a college freshman who didn't know everything about trigonometry on her first day of class, there is no problem with one's energetic field being less than 100 percent vibrant and clear. Still, the freshman will benefit from learning, just as the energy field will benefit from being cleared.

Overcome Fear and Worry

Nothing protects us from all forms of negativity like true courage and calm, and nothing opens us up to negativity like letting fear and worry run the show.

Of course some fears are necessary, such as the fear of standing in the middle of a busy freeway. Other fears are natural and not entirely unavoidable, such as the fear that something will happen to a loved one or the worry that arises when we can't immediately find our car keys. When these types of fears are in their natural state (in other words, when we do not give them more power than they're worth), they are generally experienced as mild, peripheral, and nondebilitating. What's more, for the sake of our energetic well-being, it's ideal for us to cultivate an inner state that is as light and carefree as possible.

But how? There are a number of ways to approach it; I'll list some core methods here, and we'll go into some of these in more depth in chapter 10.

> **Hand It Over:** When fears and worries come up—whether you're sleeping, eating, meditating, or working—make a conscious decision to give them over completely to the Divine. Don't try to erase them or stop worrying about them. Just mentally hand them over and then let go, as if you're handing over a dirty diaper or a large restaurant check. This way, whatever you're worried about will be handled in the most ideal way possible. If you need to take action, your intuition and inner knowing will let you know. Other times, your divine helper(s) will have it covered.
>
> When you hand fears over in this way, you are performing alchemy. In other words, you're consciously transmuting the negative energy generated by fear into the most positive energy possible: divine wisdom and power.
>
> **Meditate:** When you stick to a daily (or close to daily) meditation of at least ten minutes, during which you consciously tune in to divine energy (i.e., the Infinite, the Eternal, the world between the worlds), it is only a matter of time before you find that you are grounded in a powerful feeling of trust, confidence, courage,

and calm. The meditation practice that we've already discussed would work perfectly for this purpose.

Exercise: Nervous energy can live in our bodies. Exercise transmutes nervous energy into relaxation and calm. All exercise can work for this purpose, but for most energy workers, a conscientious-movement practice (such as yoga, tai chi, or qigong) coupled with aerobic exercise works best. For example, you might do yoga or tai chi three days per week, jog three days per week, and rest one day per week. Try to go for at least twenty minutes per day, but go longer if you can and when you feel inspired. Also, *some* exercise is always better than *no* exercise, so if all you can convince yourself to do is five minutes' worth, that's great too!

Make the Decision: Sometimes all you need to do is make the decision to rise above fear and worry. Put your foot down. Live in your power and your strength. Just say no to fear and worry. You don't have time for it, and it's not doing anyone any good. Choose in this moment to live boldly and courageously, and give fear and worry the old heave-ho.

Say Affirmations: If you find that your mind is riddled with fear and worry, you might choose an affirmation and inwardly repeat it over and over again to keep your mind occupied and replace the negative vibes with positive ones. For example, you might inwardly repeat this: "My consciousness is now flooded with the divine light of love. I am safe, and all is well." Or you might repeat something simple, like "Everything is perfectly unfolding." When you notice your mind wandering, it's no problem! Just bring it back to the affirmation.

Or you might have an affirmation or two in your back pocket to pull out whenever worry strikes. For example, when I feel the need, I like to inwardly repeat this affirmation, inspired by the work of author Louise Hay: "In this moment, all is perfect, whole, and complete. Everything is working out for my benefit. I am safe."

Pray and Invoke: Saying a quick prayer or invocation can help support you in handing over your fears and worries to the Divine (see "Hand It Over," page 146) and can quickly tap you into the source of all wisdom and all knowing that resides within you. That way, you can act from your divinity rather than from your panic!

Work with Vibrational Medicine: For chronic or recurring fears and worries, flower and gem essences, as well as gemstones themselves, can be invaluable. Here are some vibrational remedies and what they help with.

RESCUE REMEDY: General stress

WHITE CHESTNUT ESSENCE: Overactive mind, excess worry

ASPEN ESSENCE: General fear

RED CHESTNUT ESSENCE: Fear about other people's well-being

Iron Pyrite Essence or Gem: Restlessness and feeling unsupported

Moss Agate Essence or Gem: Inability to trust that everything is perfectly unfolding

Jasper Essence or Gem: Fear and worry stemming from lack of self-love

nine

BLESSING AND
FINE-TUNING THE SPACE

magine there's a song running through your head, and it's driving you crazy. In order to stop having that song running through your head, we all know that you can't merely decide to stop thinking about the song. You must simultaneously decide to think of a different song or think of something else altogether, otherwise the annoying song will probably just show right back up.

Blessing and fine-tuning the space is like helping your space to hum a different tune, infusing the space (and therefore your life) with the fabulous vibes of the tune, and, in the process, making sure the old tune has absolutely no inclination to reappear.

Sounds fun, right? It is.

Quick and Basic Method

On its own, the quick and basic method (which I've already alluded to in chapter 6 as part of the thorough space-clearing ritual) is a wonderful way to bless and fine-tune the space in a pinch, but it's also an invaluable addition to more involved blessing rituals. Here's how you do it:

Sit or stand in a powerful spot in the house—this could be a central location, just inside the front door, or somewhere else that feels right. Hold your hands in prayer pose, close your eyes, and take some deep breaths. Visualize a very bright sphere of golden-white light completely filling and encompassing the house, gently spinning in a clockwise direction. Think or say the words "harmony and happiness," and mentally send their resonance throughout the sphere of light. (You can also choose other words according to your intention, such as "romance," "abundance," "peace," etc.)

You might also like to incorporate the quick and basic method into your daily meditation practice as a general maintenance measure.

Variations on the Quick and Basic Method

While there are a number of variations on the quick and basic method, they're still quick! No matter which variation you choose (if any), the whole thing doesn't need to take longer than a minute or two.

Of course, if it takes longer at first for you to get the visualizations down, don't worry! If you stick to it, it'll all become old hat in no time.

Energetic Envisionings

Instead of (or in addition to) thinking or stating the words, you might envision energetic qualities that infuse the space with these intentions. For example, in addition to thinking the word "prosperity," you might envision green or gold sparkles filling the sphere of white light or moving through it like an eternal fountain. For romance, you might envision etheric red and/or pink rose petals raining down and filling the sphere with their fresh romantic vibrations. (See table 9-1 on the next page for more ideas.)

Divine Invocations

Instead of (or in addition to) saying or thinking the words and/or envisioning special energetic conditions, you might invoke a divine helper or helpers to assist with your blessing and fine-tuning efforts.

It should be noted that now that we're calling in energies rather than casting them out, the divine helper you call on doesn't have to specialize in energy healing or space clearing. You might call on your favorite old standby helper or choose one or more helpers that are known for the specific qualities and energies you'd like to call in.

Beginning on page 156, you'll find some ideas for helpers you can call on for various purposes, along with some suggestions for how to invoke them. But don't feel confined to these! As always, I recommend that you work with whomever you feel drawn to work with, and call on them in whatever way feels most powerful for you. (Remember—we are in the age of spiritual creativity!)

TABLE 9-1: **Visualized Energy Infusion Ideas for the Quick and Basic Method**

ROMANCE	Pink, red, and/or white rose (etheric) petals Sparkly pink light Sparkly white (etheric) sugar
PROSPERITY	Sparkly golden light Sparkly green light (Etheric) hundred-dollar bills (Etheric) golden coins Any or all of the above moving through the home like a giant fountain
HARMONY	A cozy, warm orange glow Rainbow sparkles Rainbow light
HAPPINESS & JOY	Celebratory bright red light Bright, sparkly blue light Rainbow light Sunshine-yellow light
PEACE & SERENITY	Cool blue light Cool green light Soft peach light (Etheric) ocean waves gently moving through your home

SUCCESS	Vibrant red light
	Vibrant royal blue light
	Bright golden light
BALANCE	Clear light with rainbow sparkles
	Bright purple or indigo flame
HEALTH	Vibrant green light
	Sparkly white light
	Cool, clear water from an (etheric) well with miraculous healing properties
	Any or all of the above moving through the home like a giant fountain
CREATIVITY	Rainbow sparkles
	Rainbow light
	Pastel pink light
	Pale yellow light
	Mint green light
	Nighttime starlight
	Moonlight
FUN	Bright red light
	Vibrant rainbow light
	Sparkly rainbow light
	Pink champagne light with sparkly (etheric) upwardly moving bubbles
	(Etheric) sound of laughter

LIST OF DIVINE ASSISTANTS

Lakshmi: Lakshmi is the prosperity and luxury goddess of the Hindu pantheon. Call on her to help infuse your space with the energy of wealth and abundance. To do so, you might say "Lakshmi, I call on you!" three times, and then mentally conjure up the sound and vision of rushing water and jingling coins.

Saint Germain: Saint Germain is an ascended master who helps with alchemy and balance. Call on him to bless or fine-tune your space with his balancing, all-purpose violet flame.

Archangel Michael: By now I'm assuming you know Archangel Michael pretty well. You can call on him to bless your space with very clear, positive vibrations and the energies of courage and trust. All you need to do is inwardly or aloud say something like this: "Archangel Michael, I call on you! Please bless this space and infuse it with vibrations of courage and trust."

Archangel Raphael: The archangel associated with vibrant health. To enhance the health of your home's occupants, say a simple invocation such as this: "Archangel Raphael, I call on you! Please infuse this space with your vibrant green light of health and wellness."

Archangel Jophiel: The feminine archangel associated with balance and beauty. Call on her to infuse your space with light violet/pink light that brings everything into harmony and increases the experience of sensuality and aesthetics in your home.

Archangel Metatron: The archangel that helps with time management, organization, and focus. Call on him to help you with any or all of these, and visualize/imagine/feel his highly clarifying presence filling the space.

Saint Francis of Assisi: The Catholic saint characterized by kindness, connection to animals, and connection to all nature. Simply call on him and ask him to help protect and support your animal companions and/or to infuse your space with the sort of inspired calm that comes from spending time in a beautiful natural setting.

Saint Thérèse of Lisieux: The Catholic saint known as "the Little Flower" who taught that it's the "little way" that brings the most satisfaction and joy. If you feel depressed, overly stressed, disempowered, beaten down by life, or just like you need a fresh perspective, call on Saint Thérèse to infuse your home with the joy and humility that comes from doing little things with great love.

The Fairies: As whimsical, otherworldly spirits of nature, the fairies can swirl joy into our lives and infuse our homes with bright sparkles, fresh ideas, and inspired creativity. Call on them by saying this: "Fairies, I call on you and invite you into this space!" You may then request that they help you to lighten up, open your heart to romance, and inspire your creative projects. (PS: Fairies respond well to presents, so you might leave them a gift on your altar or outside your front door, such as a sparkly crystal or a bit of chocolate.)

Merlin: As we discussed in chapter 3, Merlin, the shapeshifting magician of the Arthurian legends, is a master alchemist who can powerfully balance and fine-tune the vibes of your home with his clear light, vibrant movement, and rainbow glimmers. Call on him to balance, activate, and align the energy of your home in an ideal and masterful way. Simply say or think something like this: "Merlin, please help fine-tune the energy of the space," and then envision him vigorously and decisively stamping his staff on the floor, sending vortices of clear rainbow light into the farthest reaches and darkest corners of your home, balancing, swirling, and activating the energy in exactly the way that is most needed.

HOME-BLESSING RITUALS

When you move into a new home (after performing a thorough space clearing), or at other times when it feels appropriate to really make a point of establishing positive vibrations in your space, you might like to perform one of the home-blessing rituals below.

At different times in your life, different home-blessing rituals will probably appeal to you more than others. Each time you choose to perform one, simply choose according to what intuitively feels right and what matches your present intentions and purposes. (And, once you feel comfortable with the process, you might also create your own ritual or alter an existing one to suit you—see appendix A for ingredient ideas.)

Please note that it's best to perform these blessing rituals when the home is both physically and energetically clean. These also work for businesses and other structures.

Rosemary Home Blessing

This is an excellent, all-purpose general blessing ritual.

YOU WILL NEED:

> 5 fresh rosemary sprigs (approximately
> 4–7 inches long) for each room and area
>
> 1 18-inch length of hemp twine for each room
> and area
>
> 1 small nail for each room and area (and hammer)
> or 1 tack for each room and area
>
> A chair, small ladder, or stool (so you can reach
> the ceiling)
>
> 40 drops rosemary essential oil
>
> Water in a mister

Tie five rosemary sprigs together tightly with the hemp twine, and repeat for each bundle of five. Then hold your hands over all the bundles, palms facing the rosemary, and visualize vibrant white light coming down from above, entering the crown of your head, going down to your heart, then out your palms and into the herbs. Say:

I give thanks to these plants for being here
and to the earth for providing them.

May they bless this home and infuse its every molecule
with the energies of wisdom, comfort, health,
happiness, and divine good fortune.

Thank you, thank you, thank you.
Blessed be. And so it is.

With the nail or tack (or in any other way that works), suspend one bundle from the center of the room. (It doesn't matter how low or high they are suspended—do whatever feels right and looks best to you.) Repeat in each room and area.

Now, put forty drops of rosemary essential oil in the mister with the water and shake. Move through the room clockwise, and lightly mist the space while repeating:

Wisdom, comfort, health, happiness,
and divine good fortune.

Repeat in each room and area.

Seal your efforts by finishing with a chosen incarnation of the quick and basic method (see page 152).

Leave the rosemary bundles up for six days, then either bury them, put them in with a fire in your fireplace, release them in a moving body of water, or place them in a compost heap. (Feel free to reuse the hemp twine.) To reinforce the effects of the ritual and experience the aromatherapeutic benefits of rosemary, refresh the space with the mist as desired.

Minty Fresh Home Blessing

This ritual brings a very scintillating and uplifting vibrancy to a space. This might come in especially handy to set a new tone after a physical or mental illness, a death in the house, or some other sort of heavy energetic condition. It would also be great to signify and set in motion a fresh new start in your life.

YOU WILL NEED:

> 1 neutral-colored (off-white) soy votive candle for each room and area
>
> 1 glass or ceramic candleholder for each room and area
>
> Enough fresh mint leaves (any variety) so you'll have one small handful for each room and area, all placed in a bowl
>
> Mortar and pestle
>
> Peppermint essential oil
>
> Spearmint essential oil (optional)
>
> A mister of water
>
> Matches or a lighter

Assemble the ingredients. Stand near them as you center yourself by relaxing and taking some deep breaths. Hold the bowl of fresh mint leaves in both hands, and say:

> *For these fresh leaves, I give thanks to the earth*
> *and the powers that be.*

Place one small handful in the mortar and lightly crush them with the pestle. As you do so, say:

*I now release the magical essence of these leaves,
and I request and affirm that they now permeate this home
and bless this space with their sweet freshness.*

Transfer the crushed leaves to one of the candleholders. Using your finger, lightly coat the entire surface of the candle—excluding the wick—with the peppermint oil. (Skip this step or use a glove if your skin might be sensitive to the oil.) Place the candle in the center of the candleholder, with the base on top of and surrounded by the leaves. Put the candle, candleholder, and leaves in a central location in the room, and light the candle.

Take the bowl of leaves and the mortar and pestle, along with one candle and one candleholder, to the next room or area. Put another small handful of leaves in the mortar, and repeat the above process. Repeat in each room and area.

Return to your starting location and place twenty drops of peppermint oil and twenty drops of spearmint oil (or just forty drops of peppermint oil) in the mister of water. Close the lid and shake. Hold the mister in both hands and visualize very bright minty-green light with white sparkles coming down from above, entering the crown of your head and going down to your heart, down your arms, through your hands, and into the bottle. When this feels complete, mist the perimeter of the room in a clockwise direction. As you do so, envision this minty-green sparkly light filling the room. Repeat in each room and area.

Seal your efforts by standing in a central location and performing any variation of the quick and basic method (see page 152).

Allow the candles to continue to burn for at least one hour. Extinguish and place the mint leaves outside on the earth (soil) near the front door and around the perimeter of the home, or—if you don't have exposed earth near the outside of your home—place them in a compost bin or at the base of a tree somewhere. Feel free to use the candles again later for other purposes.

Sweetness and Light Home Blessing (aka Sugar Cookie Blessing)

This ritual fills your home with sweetness and warmth like no other! It also very effectively calls in abundance and enhances household harmony.

As an added bonus, it features a very fun (and delicious!) post-ritual activity that allows you and your loved ones to internalize the magic.

INGREDIENTS FOR COOKIES:

¾ cup organic sugar

¼ teaspoon cinnamon

1 cup margarine

½ banana, mashed

1 teaspoon vanilla extract

½ teaspoon salt

2¼ cups flour

¼ teaspoon baking soda

Heart-, star-, or sunshine-shaped cookie cutters

Note: This cookie recipe was adapted from *The Garden of Vegan* cookbook by Tanya Barnard and Sarah Kramer.

INGREDIENTS FOR THE REST OF THE RITUAL:

Organic sugar

Cinnamon

One coffee mug, glass, or small ceramic dish for
each room and area

One stick of cinnamon incense and one stick of
vanilla incense for each room and area

Open all the interior doors in the house (so the aroma can permeate) as you bake the cookies.

Preheat oven to 375 degrees. In a large bowl, stir together the sugar, cinnamon, margarine, banana, and vanilla. Add in the salt, flour, and baking soda, and mix. As you stir, quietly yet purposefully chant the words "sweetness, light, and happiness bright." Divide the dough in half, and roll out each half on a lightly floured surface. Close your eyes and say a quick blessing prayer in any way that feels right to you, envisioning the dough being filled with very bright white light. Cut into shapes and bake for eight to ten minutes or until edges are browned.

Allow the cookies to cool on the counter. As they're cooling, fill each cup or dish with about an inch and a half of sugar and one teaspoon of cinnamon. Mix. Then light one stick of vanilla incense and one stick of cinnamon incense, and stick them both in the middle of the cup or dish so that they're standing straight up (they shouldn't be touching each other). Place in a central location in the room you're in, and then repeat the process in each room and area.

As the incense is burning, stand in a central location and perform the quick and basic method (see page 152). Visualize a sphere of very warm, sweet light. See white sparkles (like sugar crystals) swirling around and permeating the entire space. Follow up the quick and basic method by saying this poem:

> *Sweetest sweet and brightest bright,*
> *This home is filled with pure delight.*
> *Within these walls, as time shall tell,*
> *We are safe and all is well.*

When the incense has burned down, mindfully and with gratitude dispose of the incense remains and sugar/cinnamon mixtures, and then wash the cups or put them in the dishwasher. As soon as you can, assemble the household—you could also invite really positive and well-meaning friends over—and eat the cookies. During the gathering, in any way that feels right to you—without being overly cheesy—see if you can encourage everyone to laugh and express lots of joy and love. Maybe trick them into it by putting on an irreverently funny movie or spiking their punch (just kidding about that last one…sort of). Perhaps you could even do things like sing, dance, or play music together.

Calling In Good Spirits

Inviting sweet and benevolent spirits into your home is an effective and relatively effortless way to fine-tune the energy of your space and keep all forms of negativity at bay.

And, as usual, there's more than one way to do this. The following suggestions can be performed alone or together in any combination.

Prayer/Invocation

To call good spirits into your home, all you really need to do is ask! You can simply relax and say a prayer or invocation to any benevolent being or group of beings that you like, in any way that feels right to you. For example, you might say this:

Calling all angels! Calling all beings of blessings and light!
Guides, helpers, illuminated ancestors,
and all divine beings that wish me well,
I call on you!

You are welcome here. Please enter and dwell.
Thank you for being here and for keeping the energy
in this space positive, vibrant, and bright.
Thank you, thank you, thank you.
Blessed be! And so it is.

Or you could say:

I now call on one hundred angels! Please bless and protect this
space. Please surround this space and dwell in this space.
Thank you, thank you, thank you!
Blessed be. And so it is.

(You get the picture.)

Beings you can invite include:

- Deceased loved ones and ancestors
  who have passed into the light

- Angels

- Fairies

- Divinities

- Ascended masters
- Spirit guides
- Beneficent spirits of the land

See chapter 3 for more ideas and descriptions.

Sweetgrass

Smudging with a braid of sweetgrass or burning sweetgrass incense are ways to invite sweet and benevolent spirits by sending a fragrant invitation out into the ether. To think of it another way, the smoke from sweetgrass creates a doorway of light between the worlds, through which divine and helpful beings are drawn.

If you have a braid (available online and at many health food and metaphysical stores), light it so that it's smoking like incense (not burning like a candle). Carry a dish under it to catch burning embers, and slowly walk around each room and area in a clockwise direction. If you have incense sticks or cones, simply light one and place it in an incense holder. You can place the incense holder on your altar if you have one (see below) or in a central location.

Altars

Whether it's simple or elaborate, creating an altar is a powerful way to call in the presence of a divine or benevolent being or beings into your home and to consistently encourage and express gratitude for their presence.

In addition to inviting divine beings, altars can be created to honor and invite the presence of an ancestor. For example, you might choose to create an altar to your great-grandparents or a particular forefather or foremother who you have always felt watching over you and wishing you well.

A modern way of doing this—and one that works wonderfully in small spaces—is to forgo a flat surface and create a simple ancestor/deceased-loved-one altar out of framed pictures on the wall. (My boyfriend and I have one of these "wall altars" in our house. Along with black-and-white pictures of our grandparents, we added our beloved cat Smoke's old collar, a brooch that used to belong to Ted's grandma, and my Grandma Deane's wedding ring.)

For more information on creating an altar, see chapter 3.

Offerings

When I was a child, Saint Patrick's Day was my favorite holiday. Even though no one suggested that I do this, I remember insisting on leaving out treats "for the leprechauns" the night before. My mother (who is Irish) apparently loved this idea, as illustrated by the fact that on the morning of March 17, wherever I had left the treats, I found all kinds of sparkly green goodies and a cheery, grateful little note from the leprechauns.

I suspect that the idea to leave an offering to the leprechauns wasn't merely a childish invention but a genetic or past-life memory of Old World customs. It's said that many people left milk outside the front door every night "for the fairies" (to curry their favor and entreat their assistance), and that other offerings of food and beverages might be left inside for the "house fairies" or "house brownie (spirit)."

It's not hard to imagine that the custom of leaving milk and cookies out for Santa is in many ways another carryover of this tradition.

In countless cultures and ages, people have offered things such as food, beverages, and fragrant smoke to gods, goddesses, ancestors, and spirits of the land.

In addition to placing offerings on an altar, here are some ideas for ways you might make offerings to helpful beings in order to invite their presence and request their assistance in keeping your home safe, blessed, and energetically fine-tuned. Feel free to put your own spin on these.

- Make an offering to the fairies and spirits of the land by getting a bird feeder and keeping it stocked with fresh bird food or nectar.

- Leave whole grain breadcrumbs or cooked rice out for the air spirits (and birds).

- Leave nuts out for the spirits of the earth (and squirrels).

- Place walnut shell halves filled with beer, cider, or champagne outside in the garden or yard—it is said that fairies love bubbly party beverages.

- Hang wind chimes near your front door and offer their sound to the angels.

- Light incense and offer its fragrance to God/dess.

- Pour a glass of wine or cider and place it on the dinner table as an offering to the earth. After dinner, go outside and pour it near the base of a tree.

- Plant and tend an organic garden or outdoor potted plants to summon fairies (and also ladybugs, birds, bees, etc.).

- Bring potted plants into the house to call in nature spirits.

Incense, Essential Oils, Crystals, Flowers, and Mist Potions

Additionally, vibrational items and ingredients such as incense, essential oils, crystals, flowers, and mist potions can be employed to energetically fine-tune your space. For a list of these types of ingredients, see appendix A.

ten

Good Habits and Spiritual Maintenance

ow that we've reached the final chapter of the book, I'd like to reiterate some key points:

- We dwell in a sea of energy.
- Our thoughts, feelings, and intentions powerfully affect this sea of energy.
- Everything is connected.
- Like attracts like.

With this in mind, it's easy to see that our spiritual habits, recurring thoughts, and general state of mind affect every aspect of our lives, including the energy of our homes and workspaces. So, like tending a garden, energy clearing and spiritual maintenance are ongoing (and rewarding!) pursuits.

Because you've now undergone an initial shift (chapter 2), and are maintaining good magical hygiene practices (chapter 4), you're off to a great start. This chapter is about deepening your practice, honing your skills, and keeping up a positive energetic momentum.

Cultivate Positivity

What we focus on expands—and thoughts, feelings, and beliefs can easily become habits. Additionally, when our minds are clear and positive, it's easier for us to access our intuition and spiritual power. That's why it's important to notice our internal monologues and shift them in ways that lift our spirits and, in the process, help us experience more of what we want and less of what we don't.

And this is a continuing practice! You might think of it this way: life is one big art class. The universe is the teacher, your life experience is your canvas, and your thoughts, feelings, and emotions are your paints. (But you're still a student, remember, so have fun with it—there's no need to take anything too seriously.)

When it comes to cultivating positivity, there are a million different teachers, a million different books, and a million different techniques all aiming at the same target. Here's a small sampling of some things that I've found work well for me. Try these or find your own!

Yoga and Breathing Exercises

Now that I've established the habit of an almost-daily yoga practice, I can't imagine staying positive without it. Whether I do ten minutes or ninety minutes, a yoga session calibrates my mood, enhances my intuition, strengthens my focus, and enlivens my spirit like nothing else.

Breathing exercises, or *pranayam*, done alone or in tandem with a yoga practice activate vibrant energy and clear stagnant energy from the body, mind, and emotions.

Everyone's different, so it's important to find a form of yoga or conscious breathing that works for you—or perhaps find another form of conscientious, meditative exercise altogether, such as tai chi or qigong.

Spend Time in Nature

For most of the time that humans have been on this planet, we've spent most of our time outdoors. Outdoors, our awareness naturally expands and becomes more inclusive. There is movement all around us and above us. The sky stretches on forever, the breeze caresses our skin at varying temperatures and from various directions, magical creatures surprise and bless us with their presence, and we inhale ever-shifting fresh scents carried along by the fresh air. In beautiful outdoor settings, we gain a truer perspective, and it feels natural to know that we are at one with everything.

I do my best to spend quality time outdoors at least once a week by hiking in the woods, jogging along the beach, watching the sunset, or sitting meditatively in a park or botanical garden. But just sitting on your balcony or in your backyard and gazing at the clouds, petting your dog, or listening to the sound of the wind in the trees can have the same effect.

Forgive

When we hold on to grudges, grievances, or old hurts, we are giving away our power. This doesn't mean that we should just say "that's okay" if someone has mistreated us. It just means that if, in our own minds, we define or limit ourselves by that mistreatment, we are giving power to the person who has mistreated us.

As an example, let's use my former stepfather, who molested me as a child. Imagine you were in the same room with the two of us, and imagine I said to him: "Because you did this to me, I am forever scarred, and I'll never be as happy and healthy as I would have been otherwise." Wouldn't you perceive my former stepfather as a powerful figure in my life? And wouldn't you perceive me as less powerful? On the other hand, imagine I said: "I know you did this terrible thing to me, but I forgive you. Because I have forgiven you, I have been able to heal, and I am even happier and healthier than I would have been otherwise."

Now who's the powerful one?

As I've heard the author Denise Linn say on her radio show, "You can only forgive when you're ready to forgive." So don't force it, but do become aware of people or situations that you are still harboring a grudge toward so that you can hold the intention to forgive them. To open your mind to the process and get the ball rolling, you might inwardly state or repeat, "I am willing to forgive _____ for _____."

For extra support and ideas on how to approach the forgiveness process, you might read *If I Can Forgive, So Can You* by Denise Linn or *I Need Your Love—Is That True?* by Byron Katie. There is also a helpful little forgiveness ritual called "Keys to Freedom" in chapter 1 of *Magical Housekeeping*.

This is, of course, an ongoing practice. We're pretty complex little bugs, so we can't do it all overnight, and new issues will arise. (Otherwise the art class would be over!) But we can get so comfortable with the process that we get excited when we discover we have something else to forgive. "Oh, goody," we can say, "more power to reclaim!"

Question and Let Go of Limiting Beliefs

Our minds also harbor little hidden caches of positivity and power behind our limiting beliefs. Whether we picked up these beliefs from our culture, our family, our friends, or our experiences, all we have to do to reclaim this power is let the beliefs go or replace them with more powerful ones.

We do this by first noticing the beliefs and then earnestly evaluating and questioning them. You sometimes have to do a little extra internal sleuthing to discover them, because when you've been mistaking them for cold, hard reality for some time, they tend to blend in.

To illustrate, here are a few of the many untrue limiting beliefs that I've noticed within myself in the past:

- You can't trust men.

- It's comforting to be worried about money because it's what I'm used to and what I was raised with.

- I'm not beautiful and I'm not really a valuable person because I don't look like the cover model on a magazine.

The fun thing is this: our thoughts and beliefs create our reality, so when you notice that a limiting belief doesn't have to be true, you get to weed its influence out of your life.

So, if you're like I've been in the past, and you hold on to the belief "you can't trust men," you'll see reasons to mistrust all the men in your life. Consequently, some or all of the men in your life will sense that you mistrust them for no apparent reason and will likely resent you for it. Because of this resentment, they will be more liable to behave in ways that validate your belief—and the cycle of negativity will continue and gain momentum.

If, on the other hand, you allow each relationship with each male in your life to be what it is and not be contaminated by your limiting beliefs, you will not only treat men better, you will also be more apt to notice your intuitive hits about who to trust and who not to trust, and you will naturally gravitate toward more trustworthy men.

Can you see how subtly this can all play out? That's why it's important to work hard to continually notice, question, and uproot the limiting beliefs that hold negative patterns in place in our lives.

Here's a method that works for me (I learned it from my friend Karynne Boese, a miraculous metaphysical life coach and gift revealer):

When you encounter a challenging situation that seems to be recurring in your life, such as a certain type of money, career, or relationship issue, sit down with a notebook and a pen. Then brainstorm all the limiting beliefs that you have about the issue.

Let's say you're frustrated at work, and you've been frustrated for some time because you don't feel satisfied with what you're doing and you don't know what direction to take. You might take a moment to ask yourself this: "What limiting beliefs am I holding on to around this issue?" Then you might get really honest with yourself and write:

- You can either be financially secure or creatively satisfied, but not both.

- Bosses are always jerks.

- I'm nothing special, so my job is nothing special.

- This is just the way it goes.

Then take some time with each belief, and brainstorm ways that it's actually not true. For example, to disprove "you can either be financially secure or creatively satisfied, but not both," you might think of financially successful artists or entrepreneurs whose work you admire. Then, to disprove "bosses are always jerks," you might think of friends you have who love their bosses or people you know who are bosses and who are actually really nice. With "I'm nothing special, so my job is nothing special," think of all the times that you did something that you were proud of and all the compliments you received that made you feel good about yourself. In other words, do everything you can to undermine and effectively invalidate these beliefs.

When this feels complete, revise each belief so that it more accurately portrays not only a more honest and accurate "reality" (which is a non-fixed concept to begin with, according to theoretical physicists Stephen Hawking and Leonard Mlodinow), but the reality that you would like to experience. So your new list may look like this:

- Many people are both creatively satisfied and financially secure at the same time.

· There are a lot of bosses who are actually pretty cool.

· I am unique and gifted, and I deserve to get paid well for expressing my talents.

· There is no one "way it goes." I am free to create a joyful life.

You can engage in this process anytime you notice a recurring negative pattern arise. Don't be down on yourself if the same issue comes up more than once. Be patient, and know that each time you clear your beliefs around the issue, you are healing more and more deeply and stepping into greater and greater levels of positivity and spiritual power.

Journal

Ever since I read *The Artist's Way* by Julia Cameron years ago, I've done the "morning pages" exercise almost every morning. This entails rapidly scribbling three notebook pages' worth of whatever words pop into my head, regardless of how boring, immature, or nonsensical they may be. But since long before that (second grade, actually), I've carried a journal and made lists of future goals, brainstormed ideas for projects, vented about things that made me angry or afraid, and confessed my secret fantasies, crushes, desires, and dreams.

I can't possibly express how helpful this has been and continues to be. It reveals important information about my internal environment that I otherwise might not consciously notice, and it powerfully attunes me to my inner dialogue, not just while I'm journaling but throughout the day too. Needless to say, this links me right up to my intuition and innate spiritual compass.

Give it a try! You could do the morning pages or something a little less structured. The important thing is that you give yourself the time and space to express your true thoughts and feelings. After all, how can you release a limiting belief if you don't know what you believe, how can you heal old hurts if you've forgotten you're still carrying them around, and how can you move toward your dreams if you don't know what they are?

Affirmations

If you've read much about the evolution of positive thinking over the past hundred years, you've probably read about affirmations: those optimistic little sentences that you repeat, inwardly or aloud, such as "Everything is perfectly unfolding and all is well" or "I constantly receive abundance from the Infinite Source." Yes, they're old-fashioned, but there's a reason why so many positive-thinking gurus extol their virtues: they work!

If, for example, you find that your mind keeps drifting to worrisome thoughts—say a particular money concern—with just a little bit of effort, you can turn the tide of your thoughts (and hence the tide of your life) with an affirmation. In this situation, you could begin to inwardly repeat something like, "I dwell in abundance, and the universe provides. I am safe, and all is well." When you realize that your mind has drifted back to the worry, you can then—simply, gently, and without judgment—take yourself out of the cycle of worry by again repeating the affirmation.

Another way you can employ affirmations is by choosing one or two every morning and writing them over and over again across a page or two or repeating them aloud for a predetermined length of time. This can set the tone for your day and help keep up your positive momentum. If you want to multitask, you might add affirmations to another daily task, such as washing the dishes or going for a jog.

My favorite affirmation book of all time is *The Wisdom of Florence Scovel Shinn*. I also like *You Can Heal Your Life* by Louise Hay. But there are plenty of other books containing affirmations, and you can also make up your own.

Here are a few I've composed over the years. If they resonate with you, you might start with these:

- Everything is perfectly unfolding.

- I am a clear channel of healing energy and love.

- I live from my true self and only interact with the true selves of others.

- I dance with the Divine and weave magic and beauty from the fabric of the stars.

- I constantly receive abundance from the Infinite Source.

- I am flooded with divine love—I am flooded with the light of the Goddess.

- I am a divine success in every way.

- I am balanced, vibrant, energized, and clear.

- I am healthy, happy, and free to be me.

- I embody the Goddess. I am calm, I am radiant, and I am beautiful in every way.

- I hear my intuition loud and clear—I trust my hunches, and I know just what to do.

- I relax and allow good fortune to come to me.

- My wealth is endless, and my gratitude is vast.

- Blessings are everywhere for me now, and I open my arms to happily receive them.
- I am awake to the magic of life.

Deepen Your Perception of the Subtle Realm

As you continue along this path you've started—magical hygiene, space clearing, cultivating positivity, etc.—you'll find that your perception of the subtle realm will naturally continue to deepen.

To illustrate, imagine that, for whatever reason, you'd never listened to pop music of any kind, and then someone switched on the radio. At first, no matter what song you heard—whether it was old or new, rock, pop, country, or R&B, it would all sound the same to you, with perhaps very slight variations. But the more you conscientiously listened, the more you would begin to differentiate. Eventually you'd not only be able to name the style of music but the artist, the title, and the time period too.

It's the same with the subtle realm. Now that you're consciously working with and spending time in the subtle realm, your ability to recognize invisible patterns and underlying energetic conditions is going to get a natural boost. In other words, your psychic abilities are going to increasingly awaken. You can help this process along (and make it more pleasant) by working with the following information.

Trust

The surest way to begin to get your intuitive footing is to trust what you experience. For example, as you're speaking to your brother, you might suddenly receive a vision of his dog, along with an impulse to ask him how his dog is doing. Or, as you're driving, you might hear a voice with your inner ear or in your mind that says "turn left." Whatever your impulse may be, do not discount it; acknowledge it and honor it. You may not immediately understand it in the usual sense, but that doesn't mean that it isn't valid. The more you acknowledge and trust these intuitive hits, the more they will begin to make some manner of sense to you. Which brings us to…

Don't Expect the Sixth Sense to Be Like the Other Five

Generally speaking, we expect the usual five senses to receive sensory input about tangible things, and we translate this input into meanings that most people can (at least in large part) agree upon. For example, if you discovered a note posted on your friend's front door that said "Gone to the neighbor's to feed the cat—be back soon," you would know exactly what it meant. If, on the other hand, you were performing a clearing ritual on your home and you suddenly smelled cigarette smoke (with no apparent physical explanation), you could be sensing an earth-bound entity that's been hanging around, you could be sensing predecessor energy (i.e., leftover vibes from a previous resident), or you could be sensing your beloved, beneficent, crossed-over-into-the-light grandmother, who's letting you know she's around and that she's helping with the clearing.

So—how would you tell which it was? One way would be to tune in to your emotions and physical sensations as you smelled the smoke. Did you feel uneasy, angry, uncomfortable, or stifled? Or did you feel comforted, confident, energized, or loved? You might also ask your divine helpers—silently or aloud—"Why am

I smelling cigarette smoke?" Then you can tune in to the psychic information you receive. At first you may feel like you're fumbling around in the dark just a bit, but in time you'll become more and more comfortable with the unique mechanism that we call the sixth sense.

Everyone's Psychic Abilities Are Different

I know I already mentioned this, but it bears revisiting. Just as we all have different combinations and expressions of artistic, athletic, and scholarly talents, we all have different talents for perceiving the subtle realm. To illustrate, here are some of the various ways that people receive psychic information. You might find that you receive information in all of these ways or just one or two of them. Or, if you're like me, you might start with one or two and eventually branch out into all of them. (Alternatively or additionally, you might receive information in a different way altogether—perhaps even in a way that is unique to you.)

Touching Objects: You may touch objects and receive information about the objects and/or the people who own them or have touched them in the past.

Scent: You may receive information through scents, such as the perfume of a deceased loved one or a musty smell to signify stagnant energy or clutter.

Inner Knowing: You may find that you "just know" things.

Image Flashes: You may receive inner flashes of images. Some of these might be unique to the situation and others may be recurring to indicate specific meanings. For example, when I flash on a giraffe, I've learned that it means "stick your neck out"—in other words, go for it and take a chance on something you care about.

Seeing Auras: You may see or sense colors around people's heads and shoulders. The colors indicate the spiritual and emotional state of the person they surround.

Quality of Light: You may sense energetic conditions by observing the quality of light in an indoor or outdoor setting. For example, it might appear to you as inexplicably clearer, softer, harsher, or more muted.

Waves of Emotion: You may experience waves of emotion to indicate the emotional landscape of a person or entity, or to indicate emotions that have been felt in a particular area or around a particular situation.

Physical Sensations: You may feel physical sensations (or receive a slight mental impression of a physical sensation) such as tingling, goose bumps, or warmth.

Hearing or Sensing Words or Phrases: With your physical ears or (more commonly) in your mind, you may "hear" words or phrases.

Tuning In to Names: You may receive energetic impressions and information about someone (living or deceased) by hearing or reading their first and/or last name.

Observing Someone's Home: You may intuit information about a person or a family by looking at their home, by looking at a floor plan of their home, by discussing their home, or by tuning in to the address of their home.

Go Beyond the Fear of Death

I'd venture to say that most people, on some level, harbor a very deep-seated and almost unbearable fear of death. You may think this sounds farfetched, but when you meditate, practice good metaphysical hygiene, and bravely traverse the subtle realm on a regular basis, you will find that this fear begins to dwindle. In fact, while your natural survival instincts will remain intact (i.e., your heart will still pound and you'll run for your life if you suddenly find yourself in a dangerous situation), with continued practice, the debilitating aspects of this fear will disappear almost completely.

Why? Because the profound fear of death most people harbor is actually a profound fear of the unknown and also a profound attachment to ego. When we go into the subtle realm, what we call death is no longer as unknown as it once was, and we tap into the place that transcends ego—the place where we are all one with each other and with the universe. What's more, we become aware that we can shift and shape things according to our will, so we feel empowered and blessed rather than at the whim and mercy of fate.

The implications of this are huge. When you no longer fear death in the usual way, you free up your energy and your power like never before. Everything becomes much clearer, much lighter, much easier, and much more fun. Will you still experience challenges? Of course. But they will be infused with the light of consciousness and the breath of spirit, and you can view them as learning experiences (and even adventures!) rather than setbacks.

CREATE A MORE HARMONIOUS WORLD

You are a healer. In this sea of energy that we call existence, we healers start with ourselves. Then we heal our spaces, our loved ones, and the energetic conditions in our immediate surroundings. This sets in motion great waves of healing vibrations that move throughout the planet and the universe, enhancing, blessing, and fine-tuning everyone and everything in their wake.

This is important work, and I would like to be the first to congratulate you for being called to do it. I have every faith that you will excel at (and find joy in!) your divine mission, and that you will bless all of life with your vital energy, healthy vibrations, and infinite love.

On behalf of all of existence—the seen and the unseen, the physical and the nonphysical, the known and the unknown—*thank you.*

Acknowledgments

I would like to thank Amy Glaser, Rebecca Zins, Bill Krause, Sandra Weschcke, Ellen Lawson, Steven Pomije, Jennifer Spees, Anna Levine, Elysia Gallo, Nicole Edman, Sharon Leah, Ed Day, and everyone at Llewellyn. I would also like to thank Ted Bruner, Aron Whitehurst, Michele Bartholomew, Joel Whitehurst, Sedona Ruiz, Brandi Palechek, Ellen Dugan, Courtney Lichtermann, J. P. Pomposello, Michael Milligan, Karynne Boese, Rachel Avalon, Annie Wilder, Erika Seress, Jonathan Kirsch, Anne Niven, Terah Kathryn Collins, and the Western School of Feng Shui. Additional respects are due to the authors Denise Linn, Doreen Virtue, Byron Katie, and Mary Ann Winkowski. (And, of course, a big, neverending thanks to all of my otherworldly helpers: you know who you are.)

GLOSSARY OF
USEFUL INGREDIENTS

hile countless ingredients may be used to clear, bless, and fine-tune, here's a partial list of some that I've found to be particularly useful, along with some suggestions for further reading. This section might come in handy for personalizing any of the rituals from the book or for crafting your own.

GEMS AND GEM ESSENCES

Aquamarine: Light, energizing, fresh. Used for health, protection, and clearing.

Citrine: Positive, happy, sunshiny. Used for positivity and prosperity.

Fluorite: Clarifying, aligning, calming. Used for bringing order and harmony to a space or situation.

Iron Pyrite: Grounding, energizing. Brings wealth, success, courage, and health.

Jasper: Grounding, healing, comforting. Opens the heart.

Moss Agate: Soothing, healing. Relieves stress, tension, and worry.

Rainbow Quartz: Inspiring, uplifting. Helps open a doorway of light.

Rose Quartz: Soothing, relaxing, healing. Enhances sleep and heals the emotions.

White Quartz: Energizing, clearing. Raises vibrations, protects from and helps clear negativity.

For further reading: Love Is In the Earth by Melody, *The Essence of Healing* by Steve Johnson

FLOWER ESSENCES

Aspen: Helps heal debilitating fear and replaces it with courage and calm.

Crab Apple: Imparts a feeling of internal cleanliness, order, self-love, and self-acceptance. Also great in mist potions to help clear the space of negative emotional residue.

Garlic: Helps cut unhealthy emotional cords and banish parasitic relationships in both physical and nonphysical realms. Helps protect a space from or clear out energetically draining patterns and entities.

Hornbeam: Gets energy moving in a healthy way, lends true inspiration, and helps one to find one's inner compass.

Olive: Energizes on a deep level and heals fatigue.

Red Chestnut: Helps heal excess worry and anxiety stemming from concern about the well-being of loved ones.

Rescue Remedy: All-purpose soother and relaxer. Imparts an immediate breath of balanced calm.

Rose: Clears and uplifts the vibration, opens the heart, enhances sensuality and awareness of beauty, and shines light in the darkest places.

White Chestnut: Quiets an overthinking mind. Great for resetting the vibe in a space by clearing out old patterns.

For further reading: The Encyclopedia of Bach Flower Therapy by Mechthild Scheffer, *Flower Essence Repertory* by Patricia Kaminski and Richard Katz

Essential Oils

Cedar: Very clear and spiritual vibration. Enhances strength, serenity, and confidence, and helps us find enjoyment and comfort in solitude.

Clary Sage: Clarifying, uplifting. Replaces negativity with positivity. Gets energy moving in a positive way.

Eucalyptus: Energetic disinfectant. Clears the vibe and soothes, uplifts, and supports physical and emotional health.

Frankincense: Very clear and spiritual vibration. Disrupts negative energetic patterns and transmutes negativity into positivity. Creates an ambiance loved by angels and divinities.

Juniper: Supports physical health, restructures unhealthy energetic patterns, protects, and resets the vibe.

Lavender: Gentle, all-purpose essential oil—soothes stress, adds calm energy, heals emotions, replaces negativity with positivity, resets the vibe, imparts a sense of joy and overall well-being.

Lemon: Highly cleansing, very energizing. Balancing and recalibrating to the mind/body/spirit, as it holds the energy of both sun (male) and moon (female) energies.

Myrrh: Healing, blessing, soothing, grounding. Supports health. Summons/represents divine feminine (Goddess) energies.

Peppermint: Fresh, vibrant, healing, clear. Purifies vibrations and activates ideal energetic flow.

Rosemary: Energizing, invigorating, healing. Enhances focus and memory. Activates and summons divine blessings of all kinds.

Sage: Clarifying. Imparts ancient wisdom, prosperity, and blessings while clearing and fine-tuning the vibe.

Sandalwood: Blesses and summons sweetness, sensuality, and warmth.

Spearmint: Clear, fresh vibration. Blesses the space and enhances prosperity and positivity.

Spruce: Cleanliness, order, protection, healthy energy.

Tea Tree: Lifts spirits, clears negativity, energetically disinfects.

For further reading: Magical Aromatherapy by Scott Cunningham, *Aromatherapy for Everyone* by P. J. Pierson and Mary Shipley

Incense and Smudgeables

Amber: Blesses, soothes, lifts vibration, and enchants spirits lightward.

Cedar: Lifts the vibe and strengthens, clears negativity, and imparts courage and clarity.

Cinnamon: Lifts the vibe and brings prosperity, sweetness, lightness, joy, and emotional warmth.

Copal: Opens a doorway of light and entices earthbound spirits to cross over. Enhances intuitive abilities and brings comfort with the otherworld.

Frankincense: Lifts and clears the vibe, summons angels and divinities, and dissolves heavy/stagnant energetic conditions.

Sage, Desert: Moves energy around in a healthy way and comforts, soothes, summons helpful ancestors, and imparts feelings of coziness and playfulness.

Sage, White: Lifts and clears, dissolves negative/stagnant energetic patterns, heals vibration so that only positivity remains.

Sandalwood: Blesses, enhances sensuality and romance, entices earthbound entities toward the light.

Sweetgrass: Comforts, soothes, and summons sweet and helpful spirits and beings of light. Helps recently deceased loved ones to cross over into the light.

Vanilla: Brings sweetness, romance, happiness, emotional warmth, and general positivity.

For further reading: Magical Housekeeping by Tess Whitehurst

HERBS AND PLANTS

Beans: When charged with sunlight and positive intentions, beans can prevent earthbound spirits from entering, convince them to leave, and/or compel them to cross over into the light.

Cayenne: Activates and adds fiery protection.

Cinnamon: Blesses with sweetness, warmth, prosperity, and general positivity.

Garlic: Dissuades/clears parasitic entities and dispels general negativity.

Lemon: Clarifies and balances the vibration.

Mint: Imparts freshness and vibrant positivity. Enhances prosperity and romance.

Potato: Good to eat after a ritual to help ground your energy and bring you back to the physical plane.

Rose: Said to have the highest vibration of any living thing. Imparts freshness, romance, self-love, clarity, and general positivity.

Rosemary: Blesses, protects, and banishes negative or stagnant energy and entities.

Whole Grains: Good to eat after a ritual to help ground your energy and bring you back to the physical plane.

For further reading: Cunningham's Encyclopedia of Magical Herbs by Scott Cunningham

Other Ingredients

Rose Water: Great base for mist potions, as roses have a very clear and positive vibration. (Look for rose water misters in stores like Whole Foods or online; I like the Heritage Products brand atomizer, available on Amazon.com.)

Sea Salt: Absorbs and neutralizes negativity.

Sunlight: Clears negativity, energetically disinfects, and empowers items and people with positivity.

Table Salt: Similar to sea salt but more grounding and symbolic of the earth element.

Water: Good base for mist potions with essences and/or essential oils. Add sea salt for a cleansing potion. Also good to drink in abundance during and after cleansing rituals of all kinds to help move out physical toxins.

White Vinegar: Can be used as a window, mirror, and glass cleaner that employs the added benefit of clearing the energy of these items. Can also be added to things like laundry and bathwater for extra energetic cleansing and clearing action.

Mist Potions

Clear Vibe/Quiet Mind: A mister of rose water with four drops white chestnut essence, four drops crab apple essence, and fifteen drops essential oil of cedar.

General Well-Being and Stress Relief: A mister of rose water with four drops Rescue Remedy, ten drops essential oil of peppermint, and ten drops essential oil of lavender.

Protect/Enliven: A mister of water with four drops garlic flower essence, two drops hornbeam essence, ten drops essential oil of juniper, and five drops essential oil of spruce.

Relationship Harmony and Healing: A mister of rose water with four drops rose essence and twenty drops essential oil of lavender.

Rosemary Blessing: A mister of water with forty drops essential oil of rosemary.

Wealth: A mister of water with five drops citrine essence, ten drops essential oil of spearmint, and ten drops essential oil of sage.

For further reading: Magical Housekeeping by Tess Whitehurst

appendix b

GOOD ENERGY
ESSENTIALS

n case you feel overwhelmed by all the practices presented in this book, I've designed this handy little checklist to remind you of what I perceive as the essentials: the energetic prerequisites to happy, balanced, and harmonious living. I suggest gently working your way up to incorporating each of these activities in the ways suggested.

DAILY OR ALMOST DAILY

- Enlisting help (praying/invoking divine help)
- Healthy eating
- Drinking water: at least half your body weight in ounces per day
- Blessing food/water

- Meditation:

  Clearing aura/chakras/body
  Grounding
  Connecting with sky
  Shielding aura/body
  Clearing home
  Shielding home

- Yoga or other conscientious exercise

Regularly

- Clutter clearing
- Getting in the habit of staying as clutter-free as possible
- Cleaning with intention
- Keeping energy clear with space clearing (rattles, sage, ritual, etc.)
- Fine-tuning the space with the quick and basic method

As Necessary

- Affirmations
- Flower and gem essence healing
- Special energy cures
- Advanced personal protection techniques
- Thorough clearing rituals
- Intense home-protection rituals
- Clearing objects or other people

BIBLIOGRAPHY

Anugama. *Shamanic Dream*. Open Sky Music, 2000.

Barnard, Tanya, and Sarah Kramer. *The Garden of Vegan*. Vancouver: Arsenal Pulp Press, 2002.

Brett, Ana, and Ravi Singh. *Fat Free Yoga: Lose Weight and Feel Great*. Raviana.com, 2005.

Bruce, Robert. *Astral Dynamics: The Complete Book of Out of Body Experiences*. Newburyport, MA: Hampton Roads Publishing, 2009.

Burney, Diana. *Spiritual Clearings: Sacred Practices to Release Negative Energy and Harmonize Your Life*. Berkeley, CA: North Atlantic Books, 2009.

Cameron, Julia. *The Artist's Way: A Spiritual Path to Higher Creativity*. New York: Tarcher, 1992.

Campbell, Joseph. *The Hero with a Thousand Faces*. New York: Pantheon Books, 1949.

Collins, Terah Kathryn. *The Western Guide to Feng Shui: Room by Room*. Carlsbad, CA: Hay House, 1999.

Cunningham, Scott. *Cunningham's Encyclopedia of Magical Herbs*. St. Paul, MN: Llewellyn, 1985.

———. *Magical Aromatherapy: The Power of Scent*. St. Paul, MN: Llewellyn, 1985.

Ewing, Jim Pathfinder. *Clearing: A Guide to Liberating Energies Trapped in Buildings and Lands*. Findhorn, Forres, Scotland: Findhorn Press, 2006.

Geddes, Neil, and Alicen Geddes-Ward. *Faeriecraft*. Carlsbad, CA: Hay House, 2005.

Hawking, Stephen, and Leonard Mlodinow. *The Grand Design*. New York: Bantam Books, 2010.

Hay, Louise. *You Can Heal Your Life*. Carlsbad, CA: Hay House, 1984.

Illes, Judika. *Element Encyclopedia of 5000 Spells*. London: HarperElement, 2004.

Johnson, Steve. *The Essence of Healing: A Guide to the Alaskan Flower, Gem, and Environmental Essences*. Homer, AK: Alaskan Flower Essence Project, 1996.

Kaminski, Patricia, and Richard Katz. *Flower Essence Repertory: A Comprehensive Guide to North American and English Flower Essences for Emotional and Spiritual Well-Being*. Nevada City, CA: Flower Essence Society, June 1994.

Katie, Byron, and Michael Katz. *I Need Your Love—Is That True? How to Stop Seeking Love, Approval, and Appreciation and Start Finding Them Instead*. New York: Harmony Books, 2005.

Katie, Byron, and Stephen Mitchell. *Loving What Is: Four Questions That Can Change Your Life*. New York: Three Rivers Press, 2003.

Kingston, Karen. *Creating Sacred Space with Feng Shui*. New York: Broadway Books, 1997.

Linn, Denise. *If I Can Forgive, So Can You: My Autobiography of How I Overcame My Past and Healed My Life*. Carlsbad, CA: Hay House, 2005.

———. *Past Lives, Present Miracles: The Most Empowering Book on Reincarnation You'll Ever Read ... in This Lifetime!* Carlsbad, CA: Hay House, 2008.

———. *Space Clearing A–Z: How to Use Feng Shui to Purify and Bless Your Home*. Carlsbad, CA: Hay House, 2001.

Melody. *Love Is In the Earth: A Kaleidoscope of Crystals*. Wheatridge, CO: Earth-Love Publishing House, 1995.

Pierson, P. J., and Mary Shipley. *Aromatherapy for Everyone: Discover the Scents of Health and Happiness with Essential Oils*. Garden City Park, NY: Square One Publishers, 2004.

Ramsland, Katherine. *Ghost: Investigating the Other Side*. New York: St. Martin's Press: 2001.

Rost, Amy. *Natural Healing Wisdom & Know-How: Useful Practices, Recipes, and Formulas for a Lifetime of Health*. New York: Black Dog and Leventhal Publishers, 2009.

Ruiz, Don Miguel. *The Four Agreements: A Practical Guide to Personal Freedom, A Toltec Wisdom Book.* San Rafael, CA: Amber-Allen Publishing, 1997.

Scheffer, Mechthild. *The Encyclopedia of Bach Flower Therapy.* Rochester, VT: Healing Arts Press, 1999.

Shinn, Florence Scovel. *The Wisdom of Florence Scovel Shinn.* New York: Fireside, 1989.

Shroder, Tom. *Old Souls: Compelling Evidence from Children Who Remember Past Lives.* New York: Simon and Schuster, 2001.

Starhawk. *The Spiral Dance: A Rebirth of the Ancient Religion of the Great Goddess.* New York: HarperCollins, 1979.

Tolle, Eckhart. *A New Earth: Awakening to Your Life's Purpose.* New York: Plume, 2006.

———. *The Power of Now: A Guide to Spiritual Enlightenment.* Novato, CA: New World Library, 2004.

Venolia, Carol. *Healing Environments: Your Guide to Indoor Well-Being.* Berkeley, CA: Celestial Arts, 1988.

Virtue, Doreen. *Archangels and Ascended Masters.* Carlsbad, CA: Hay House, 2003.

———. *Divine Guidance: How to Have a Dialogue with God and Your Guardian Angels.* Carlsbad, CA: Hay House, 1988.

———. *The Lightworker's Way: Awakening Your Spiritual Power to Know and to Heal.* Carlsbad, CA: Hay House, 1997.

Von Praagh, James. *Ghosts Among Us: Uncovering the Truth About the Other Side.* New York: HarperCollins, 2008.

Whitefeather, Sapokniona. *Master Meditations.* Spirit Hawk Recordings, 2004.

Whitehurst, Tess. "6 Steps to Good Magical Hygiene." *New Witch Magazine.* Issue #18.

———. "Magical Drinking Water." *Llewellyn's 2010 Magical Almanac.* Woodbury, MN: Llewellyn, 2009.

———. *Magical Housekeeping: Simple Charms and Practical Tips for Creating a Harmonious Home.* Woodbury, MN: Llewellyn, 2010.

———. *Magical Clutter-Clearing Boot Camp.* Amazon.com Kindle edition, 2011.

Wilder, Annie. *House of Spirits and Whispers: The True Story of a Haunted House.* Woodbury, MN: Llewellyn, 2005.

Winkowski, Mary Ann. *When Ghosts Speak: Understanding the World of Earthbound Spirits.* New York: Grand Central Publishing, 2007.

Wolfe, Amber. *Personal Alchemy: A Handbook of Healing and Self-Transformation.* St. Paul, MN: Llewellyn, 1995.

INDEX

GET MORE AT **LLEWELLYN.COM**

Visit us online to browse hundreds of our books and decks, plus sign up to receive our e-newsletters and exclusive online offers.

- **Free tarot readings • Spell-a-Day • Moon phases**
- **Recipes, spells, and tips • Blogs • Encyclopedia**
- **Author interviews, articles, and upcoming events**

GET SOCIAL WITH **LLEWELLYN**

Find us on
Facebook

www.Facebook.com/LlewellynBooks

Follow us on

www.Twitter.com/Llewellynbooks

GET BOOKS AT **LLEWELLYN**

LLEWELLYN ORDERING INFORMATION

Order online: Visit our website at www.llewellyn.com to select your books and place an order on our secure server.

Order by phone:
- Call toll free within the U.S. at 1-877-NEW-WRLD (1-877-639-9753)
- Call toll free within Canada at 1-866-NEW-WRLD (1-866-639-9753)
- We accept VISA, MasterCard, and American Express

Order by mail:
Send the full price of your order (MN residents add 6.875% sales tax) in U.S. funds, plus postage and handling to: Llewellyn Worldwide, 2143 Wooddale Drive Woodbury, MN 55125-2989

POSTAGE AND HANDLING

STANDARD (U.S. & Canada):
(Please allow 12 business days)
$30.00 and under, add $4.00.
$30.01 and over, FREE SHIPPING.

INTERNATIONAL ORDERS:
$16.00 for one book, plus $3.00 for each additional book.

Visit us online for more shipping options. Prices subject to change.

FREE CATALOG!

To order, call
1-877-
NEW-WRLD
ext. 8236
or visit our
website

Simple Charms
& Practical Tips for
Creating a Harmonious Home

MAGICAL
HOUSEKEEPING

Tess Whitehurst

MAGICAL HOUSEKEEPING

Simple Charms & Practical Tips
for Creating a Harmonious Home

TESS WHITEHURST

Every inch and component of your home is filled with an invisible life force and unique magical energy. *Magical Housekeeping* teaches readers how to sense, change, channel, and direct these energies to create harmony in their homes, joy in their hearts, and success in all areas of their lives.

In this engaging guide, energy consultant and teacher Tess Whitehurst shares her secrets for creating an energetically powerful and positive home. Written for those new to metaphysics as well as experienced magical practitioners, *Magical Housekeeping* will teach readers how to summon success, happiness, romance, abundance, and all the desires of the heart. And, by guiding them to make changes in both the seen and unseen worlds simultaneously, this dynamic and delightful book will help to activate and enhance readers' intuition and innate magical power.

978-0-7387-1985-6

$16.95 • 5³⁄₁₆ x 8 • 288 pp.

bibliography, appendices, index